The Choice

A STORY OF SURVIVAL

THE CHOICE

A STORY OF SURVIVAL

By Monte Anderson

ISBN: 978-193-0536852
Library of Congress Control Number: 2015955486

Best Publishing Company
631 US Highway 1, Suite 307
North Palm Beach, FL 33408

In the Sea of Cortez in 1982, David Scalia sustained an uncontrolled ascent from 110 feet beneath the surface. During his rescue and evacuation to San Diego, California, he suffered respiratory arrest. Experts in diving medicine thought his chance for survival was poor. This is the story of an incredible series of connected events, the people who directed them, and the life of a remarkable young man.

This book is dedicated to David Scalia and to each of the persons who played a part in his incredible rescue.

THE CHOICE

A STORY OF SURVIVAL

Preface

What a privilege it was to write this very special story about my friend, David Scalia. We have spent many hours skiing and mountain biking together. As we drove back from a ski trip to Colorado, he told me a short version of what had taken place in the Sea of Cortez in 1982. It was such a compelling tale, I knew others would enjoy it, too. In the writing process, I learned more about David's fundamental nature and the world of scuba diving.

When I contacted Dr. Paul Phillips to ask if he remembered the case, he answered, "I remember it perfectly well. I was the doctor who went to Mexico and picked him up."

In the entire time I have known David, he has always maintained self-discipline and calmness, traits much appreciated in the sports we share. I've never heard him curse or seen him angry. The self-confidence started in his childhood has been strengthened by his life experiences. It is clear to me the ordeal he endured was an epiphany for him, which led to a heightened sense of gratitude for his life as well as a stronger belief in a higher being.

While pondering the multiple life-threatening challenges faced by those involved in his rescue, I reflected on the possibility that we're all being judged on the choices we make in our lives, and the One making the judgment would have been pleased with the human element in David's story. At least from a human point of view, no one failed to do what he or she could. This makes every person in David's story a hero.

Since David's story took place more than thirty years ago, I was unable to contact certain individuals. John Carter, Jim O'Connor, Kyle, Mary, and Carrie Hutchings, and George Calloway are pseudonyms for real-life heroes in David's story. All other persons have provided their written consent to be included in this book.

It was my wife, Pat, who pointed out David's lovely feature — when you look into his face "it's like a smile just waiting to happen."

Monte Anderson
Prescott, Arizona
2014

San Diego

Thanksgiving Day — November 25, 1982

Dr. Paul Phillips had just completed an all-nighter in the emergency room at the University of California at San Diego (UCSD) Medical Center when he got the call. Critical Care Medicine, Inc. (CCM), an emergency evacuation service where he moonlighted in his "spare time," relayed the details of a serious situation. A twenty-eight-year-old US citizen was unresponsive after a scuba diving accident and was in a hospital near Guaymas, Mexico. A decompression chamber, the device used for definitive care in serious diving injuries, was not available in the area. The nearest chamber was located aboard a US Navy ship in port at Naval Base San Diego, home of the Pacific Fleet. Captain M.A. Heuberger had authorized treatment in the chamber aboard the navy submarine tender, USS *Dixon*, if needed.

The final plan was to evacuate the patient, David Scalia, to San Diego and transport him via helicopter to the USS *Dixon* as soon as possible. With no decompression chamber at UCSD Medical Center and no intensive care unit at Naval Base San Diego, the two facilities pooled resources, as in all difficult cases.

Convincing himself he could nap on the flight to Mexico, Dr. Phillips agreed to head to the airport for immediate departure. It would be his sixth rescue trip south of the border. He tucked the notes he had taken from the dispatcher into the breast pocket of his lab coat. Then he grabbed his briefcase, where he kept a copy of Bennett and Elliott's *The Physiology and Medicine of Diving*, the authoritative work on diving injuries, and headed for San Diego International Airport.

With a wave to the gate attendant at the private entrance to the airport, he entered the field and drove past a hundred small aircraft before reaching the CCM hangar — two prop jets inside and one outside, ready to go. The pilot and copilot were already settled in the cockpit. Nurse John Carter greeted Dr. Phillips as the young doctor maneuvered his tall, thin frame through the door of the small aircraft. The copilot pulled up the door and locked it. Dr. Phillips curled himself into one of the seats and drifted into a state of hypnotic dreams.

About three hours later, he became fully alert when the plane circled the airport. As they approached, the pilot pulled up the nose of the plane to avoid power lines inexplicably placed near the end of the runway. To Dr. Phillips, the entire airport facility felt a bit "primitive" for the 1980s.

They taxied toward a uniformed official and a man dressed in hospital whites. The plane stopped, and the engines shut down. The official motioned for the doctor and nurse to deplane, and they were promptly escorted to a waiting ambulance. Dr. Phillips grabbed a kit of emergency resuscitation equipment and headed toward the ambulance driver. After a brief greeting, he and Nurse Carter were on their way to the hospital.

"The building looked more like a clinic, a single-storied series of connected rooms," Dr. Phillips noted.

He saw the ambulance parked near a wide door where they entered the facility. He and Nurse Carter were escorted into a large room with several beds, only one occupied. A hospital nurse presented the medical notes to Dr. Phillips. Having cared for many Spanish-speaking patients in San Diego, he had learned the language fairly well, especially when it came to medical matters.

An American woman, who was seated at the bedside as they entered, stood as the rescue team approached.

"Thank God you're here," she said to Nurse Carter. She identified herself as Dorothy Centa, a friend of David Scalia, who had placed the call for help.

The nurse's response was brief. He gave her his name, said they would do everything they could, and asked her to be seated in the waiting area until they were ready to leave.

Dr. Phillips focused on the patient, shocked to observe no monitoring equipment and no Mexican doctor in sight. The patient lay unconscious, an IV catheter in place. Oxygen flowed to his nose through green plastic tubing, which was attached to a scraped-up oxygen tank standing on the floor. Dr. Phillips went into his "high-alert" mode, performing a rapid assessment. The patient wasn't responding to anything he said or did. And when Dr. Phillips lifted the patient's right eyelid, the pupil was dilated and the eye itself was helplessly deviated outward and downward.

"My God," the doctor thought, "he's already herniating."

Brain herniation is a deadly reaction to high pressure inside the skull, a rigid structure unable to expand to accommodate the swollen brain tissue. The brain tissue gets squeezed and pushed through openings in the skull such as the large hole at its bottom, the foramen magnum, where the spinal cord exits the skull and enters the vertebral column. The foramen is also where major blood vessels to the brain enter from below, and it is a life-threatening event for any of those vessels to be pinched off. In David's case, the oculomotor nerve was being compressed, causing the pupil of the eye to dilate, unable to constrict in response to light.

This nerve also controls eye movement, and when it doesn't function, the eye deviates downward and outward, as Dr. Phillips described.

He removed from his kit an endotracheal tube, a clear plastic device that he skillfully slipped through the patient's mouth and into the windpipe, giving direct access to the lungs. The windpipe end of the tube was equipped with a small ballooning collar to be inflated after insertion. The cuff sealed the trachea around the tube, allowing oxygen to be delivered by squeezing an Ambu bag (a bag valve mask) or by a mechanical respirator connected to the tube. The seal prevented oxygen intended for the lungs from escaping backward through the mouth.

Once Dr. Phillips had verified the tube was properly positioned, he rapidly pumped the Ambu bag. The coordinated rise and fall of the patient's chest in reaction to his squeezing the bag confirmed proper function. Nurse Carter listened with his stethoscope for breath sounds and gave a "thumbs-up" to the doctor. Next, Dr. Phillips gave David a dose of mannitol, a sugar alcohol that helps reduce brain pressure.

"It was one of those cases; within five minutes of arriving, he'd been intubated," Dr. Phillips said. "He was about to herniate from cerebral edema. It looked like the patient was going to pass away quickly."

Hyperventilation was critical. He concluded the Guaymas doctors had decided the case was untreatable and determined that giving the patient oxygen was all they were able to do at their facility. In their defense, it seemed clear to the doctor the only medical equipment that could help at this point was a decompression chamber.

At 7:30 p.m. Dr. Phillips said, "There didn't seem to be any [Mexican] doctors at the facility. The Life Flight team just took over."

Nurse Carter continued to pump the Ambu bag, while Dr. Phillips recorded his notes and delivered copies to the Mexican nurse along with the necessary paperwork from Life Flight. Then they moved the patient onto a stretcher and into the ambulance for the ride back to the airport. Nurse Carter escorted Dorothy to the ambulance and seated her in front with the driver. Dr. Phillips still had hardly noticed her; his focus was fixed on his patient. He directed one of the ambulance attendants to take over the pumping to continue the hyperventilation.

At the airport, the immigration official collected the departure documents and motioned for the Life Flight crew, along with the patient and Dorothy, to proceed to the aircraft. The flight was promptly cleared for takeoff, an act gratefully received because Dr. Phillips had learned firsthand that evacuation flights could be delayed, even in critical cases, by immigration officials treacherously exercising their authority. The turnaround time in Mexico was a new record for the flight crew, but Dr. Phillips was troubled as to whether or not they had arrived in time.

On the aircraft, the endotracheal tube was connected to a mechanical ventilator, and once the electrocardiograph leads were in place, the flat green line on the monitor jumped into action, tracing David's heartbeats. After settling the patient in the plane, Dr. Phillips' reassessment made him more optimistic. His patient was still alive, the pulse was regular, and the blood pressure, which was low at first, had improved. An hour later, his right eye was returning to its normal position. The young physician breathed a sigh of relief, but the challenges ahead were far from over.

"We don't hyperventilate for cerebral edema as much as we used to," Dr. Phillips reflected, "but it sure worked in this case."

Nurse Carter finally found time to address Dorothy. She sat in the rear of the plane, looking stunned but welcoming his effort to give her an update.

"We've seen some improvement," the nurse said. "We're still hoping we got here in time."

He expressed his sorrow and assured her they would be in San Diego shortly. He learned she was from Prescott, Arizona, and she had two kids. As they spoke, Dorothy put her hands over her ears, experiencing pain. John felt it as well — intense pain in both ears. Oxygen masks dropped from the ceiling. The pilot knew a loss of cabin pressure was a serious threat to the patient and shouted to the others to buckle up and hold on tight as he put the plane into a steep descent from 18,000 feet.

In the middle of the dive, the alarm on the ventilator sounded. David's face was drained of color, and everything they had achieved so far seemed to have been lost in a matter of minutes. How much more could he tolerate?

Still strapped into his seat with the obnoxious yellow mask covering his own face, Dr. Phillips struggled with the g-force and the tubing as he disconnected the ventilator and reconnected the Ambu bag. With the respirator alarm still screeching in the background, he saw that his pumping wasn't producing the expected chest movements. As the pilot pulled the plane out of the dive, Dr. Phillips was pressed tightly into his seat; it was impossible to move.

Free from the force of the aerial maneuvers, the doctor returned his full attention to David. He first checked the position of the tube, and it looked OK. He checked the air in the balloon at the far end of the tracheostomy tube. When he injected air into the port that led to the cuff, there was no resistance at all, no seal. This meant much of the oxygen he was pumping in flowed back out without reaching the lungs. Acting quickly, he removed and tossed aside the defective tube, replacing it with a new one. Thankfully, he felt the expected resistance when he injected air into the cuff. He reconnected the Ambu bag, and it worked perfectly. David's chest moved up and down with each squeeze. A few minutes later, the color returned to his face and a resilient heart resumed its rhythm. Once again, oxygenated blood had begun to perfuse David's body.

The doctor reset and reconnected the ventilator. Then he inspected the discarded endotracheal tube, finding something he had never seen before — the inflatable cuff that prevented the backflow of air was torn.

It was possible that the cuff was defective or had somehow been punctured before insertion. Another possible factor was when cabin pressure suddenly decreased, the cuff pressure, applied at sea level, would have increased significantly, expanding the cuff to the point where it ruptured. At that point, the integrity of the ventilating process would have been compromised.

Dr. Phillips closed his eyes and shook his head. He had nothing else to offer. He looked at his watch. Everything hinged on the clock. Could they get there on time?

All Nurse Carter knew was that David's heart was still beating, but nothing about whether or not his brain had been irreversibly damaged, nothing as to whether or not there could be any remedy for all the damage that had been done. It had been almost ten hours since the accident. As the lights from Ensenada, Mexico, appeared below, the pilot established contact with the control tower in San Diego and began his descent.

As the plane taxied down the runway, they saw the waiting helicopter, its interior lights shining from its open door. David was quickly transferred, and as soon as they were all seated in the helicopter, the big blades started turning. With a loud roar, they lifted off and headed for the helipad near USS *Dixon*.

It was approaching midnight when the helicopter landed. They saw several individuals standing on the ground nearby while others raced down the pier, their clothes flapping wildly from the downdraft of the rotors. The helicopter door opened, and US Navy Lieutenant Commander Gregory Adkisson and US Navy corpsman Jim O'Connor came aboard.

Dr. Phillips went over the medical history with the lieutenant commander, concluding, "We did everything we could, he may be DOA . . . we may have gotten here too late."

Akron, Ohio

1963

When David Scalia was eight years old, his father, a professional singer in Akron, Ohio, and his mother divorced. His mom often worked two jobs to provide for him and his two sisters. Left on his own for much of the time, he became independent as a young boy. He was strong, likeable, and had already developed a self-confidence many wouldn't understand. His mother was acquiescent when he stayed at a friend's house for several days; the absences evolved into hitchhiking trips. He would tell his mother he'd be back in three or four days and off he went, backpack in hand. He made the trips on his own because none of his friends enjoyed the same free rein.

"One of my favorite trips, taken when I was fourteen, was through the Smoky Mountains and the Kentucky bluegrass country," David said. "I didn't know anyone there, I just had the urge to go and see those places. They were beautiful. The next summer, I went to Florida, where I made it to the Keys and spent most of the time swimming and body surfing. I didn't meet anyone in particular, I pretty much just kept to myself."

When he was sixteen, his mother remarried and her life was easier, but David and his stepfather had a tentative relationship from the beginning.

While he was a senior in high school, David enjoyed working part-time for his uncle "Cos," who owned Scalia's AA Blueprints in Akron. Much of what he learned came from interacting with older people whom he respected. He earned enough money to move out on his own.

He graduated from high school when he was eighteen. His best friend, Kevin Sweeney, had graduated a year ahead of him. The two boys were kindred spirits. At the age of sixteen, Kevin took off on a solo trip to Europe. He left with a backpack, a Eurail Pass, and a book titled *How to See Europe on $5 a Day*. They first spoke of getting themselves to Europe by freighter, buying motorcycles, and exploring that part of the world. It was never clear where the money would come from, but both boys were optimistic about pulling it off. A security guard hired David to help with house painting, and a few months after graduation, he had $600 in his pocket. But even if the boys had pooled their resources, the trip across the Atlantic was still out of reach.

"What about Mexico?" David asked. "I've heard everything is really cheap down there, and we can learn to speak Spanish."

Kevin was in. "Let's go all the way to South America," he said. "We can hitchhike along the Pan-American Highway, and we can sleep outdoors."

The all-American hitchhiker was all in.

Whoever inspired it, both boys had become interested in philosophy, and the subject often dominated their discussions. Their backpacks contained passports, sleeping bags, and a shared Spanish-language phrase book. David added a copy of the *Dialogues of Plato*, and Kevin toted *Introduction to Aristotle.* Downtime along the many miles of roads would often be spent communing with those thinkers from 2,400 years ago.

Their departure from Akron was unceremonious. By now, David's mother was accustomed to his "excursions," and the boys left without fanfare on what would become an education.

"At first we got by for about a dollar a day most times," David said.

Rides came from all sorts of people — frequently American tourists, but often locals. David and Kevin sometimes received a lesson in Spanish from natives who seemed pleased that two young gringos were interested in their country.

"I couldn't believe how many ex-pats from the States were living down there," David recalls. "We met them everywhere."

Early on in a six-month trip, he felt a new sense of freedom he'd never experienced before. "We were really carefree, living for the moment, just taking it all in," he said.

A few years earlier, in 1953–59, Fidel Castro, leader of the Cuban revolution, revealed he was a communist. Now, much of Central and South America was involved in civil unrest or open wars. With Cuba, China, and the Soviet Union as models, communist groups formed everywhere, promising to improve the living standards of the lower classes. Fearing communism was a serious threat to their southern neighbors, the policy of the United States supported the noncommunist factions, even though many were corrupt. The result was, as it pertained to the two boys, pockets of strong anti-American sentiment along the way. The Sandinista National Liberation Front in Nicaragua had recently overthrown the dictator Anastasio Somoza Debayle. With ongoing civil war in El Salvador, the US military advisors assisted the government, while Cuba and other communist states supported the insurgents. In Peru, the militant communist group Shining Path waged their "armed struggle."

David and Kevin had heard of Castro, but otherwise the two young men learned about the rebellions only when they heard nearby gunfire along their route, which happened on more than one occasion. As they went farther south, they kept to themselves, sleeping closer to the highway. But most of the people they met on the streets and highways continued to be cordial and friendly toward the two *jóvenes.*

During their odyssey, they experienced many of the attractions that had lured them there in the first place. From the metropolis of Mexico City, they got a lift with some tourists to Teotihuacán, one of the largest ancient cities in the Americas. The two youths were amazed at the size of the ruins; this was what they had hoped to see. The enormous Pyramids of the Sun and the Moon faced one another from opposite ends of the *Calzada de los Muertos*, the Street of the Dead. More than 1,200 years ago, this was home to 200,000 people with a different set of values and beliefs.

The boys climbed the steep stairs up the Pyramid of the Sun, each stone cut and placed by hand. From the top, they watched hundreds of tourists weave their way over the walkways below. They relished the scenes and the moment.

Curious about bullfighting, they made their way back to Mexico City and to Plaza México, the largest bullring in the world. Just like Ernest Hemingway, they were fascinated by the artistry and the danger involved.

"The cruelty of the whole thing was hard to dismiss," David reflected. "*Picadores* crippled the animals before the matador used his sword."

Before leaving Mexico, they hitched their way south to the rainforests of Chiapas to see the ruins of the Maya near Palenque, a city established for 200 years before the nearby ancient ruins were discovered. These structures were much smaller but more delicate than the ones at Teotihuacán. Stone carvings from the monuments, clear as photographs, displayed the images of the Maya people from more than a thousand years ago.

Late in the afternoon, they began the hike back to Palenque, stopping for the night and hanging their hammocks between the mahogany trees. During the night, they were startled awake by a loud beastly roar in the distance. They had heard stories about black panthers in the area, and both boys were sure they had now heard one. Howler monkeys were loud and obnoxious, but this was no monkey. The two boys stayed awake for much of the night.

The next morning, they caught a ride to Salina Cruz on the west coast, and after body surfing in the late afternoon, they paid a few pesos to hang their hammocks under a *palapa*, a thatched roof shelter, for the night, a luxury they felt they had earned after the previous night in the jungle. Included in the package were generous servings of fresh fish for dinner.

"We felt like the two most blessed guys in the world," David said.

Staying along the coast for most of the day, they passed a number of rivers churning their way to the sea. Both of them had done kayaking in inflatable boats, and it was fun to mentally gauge the proper course through the rapids. But it always remained a mental game; they couldn't spend their money on such an adventure. The vote was eating 2, kayaking 0.

In Guatemala, they went to Tikal, at one time the greatest city of the Maya. Stone temples with steep, vertiginous stairways led up to ceremonial rooms on top. As in Teotihuacán, two temples stood like bookends on each end of a great plaza. Howler monkeys, hanging out in the surrounding forest, screeched their disapproval at perspiring visitors. In 1525, Hernán Cortés and his troops marched by this stunning place without seeing it for all the foliage, much of it completely covering all but the tallest structures.

Walking along the highways in El Salvador, the friends heard gunfire in the distance. They kept their heads down and hung close to the main road, finally getting another lift. It was clear that due to political positions, the locals did not like Americans. As a precaution, they decided to tell everyone they were from Canada.

In Costa Rica, they met an unlikely couple of characters. The first was a native Costa Rican, dark-complected, tall, thin and wiry. About thirty, he was bilingual and knew the territory. The other was from South Africa. A bit younger, he was more stout and had a heavy accent in English or Spanish. They invited the boys to join them for a *cerveza.* The odor of marijuana wafted from both of them. A lively conversation flowed freely, perhaps influenced by the effects of the pot.

The two new acquaintances had plotted a business venture. They were about to set off for Colombia, where, they explained, marijuana was cheap and there was plenty of it. They would buy all they could in Cartagena and smuggle the goods to Lima, where there were tons of cocaine but the wonders of "Colombian Gold" were just becoming known. They speculated the price they would get would be high . . . or they could trade pot for cocaine, ounce for ounce. When they asked the two *Americanos* if they might be interested in the venture, David and Kevin, not exactly flush with funds, were taken aback. But they didn't say no. The idea needed at least a little more thought.

The two Americans spent almost the entire night discussing whether or not to join forces in the operation. It meant becoming mules, hiding the contraband on their bodies, and successfully making it across borders in Colombia, Ecuador, and Peru. In the end, it was a risk they couldn't take.

The following day, they bought bus tickets to Panama City and left San Jose. Ruminating over the previous day, they didn't speak as they rode along, staring at the Pacific Ocean to their right.

Before reaching the town of Arraiján, mudslides over the highway brought the bus to a halt. It looked like it could be days before it was back on the road. Eyeing each other with looks of disgust, the two boys grabbed their backpacks and slogged their way to dry land.

"I've never seen so many mosquitos," David recalled. "They attacked us all afternoon."

After hiking in to Arraiján, the young men learned the road to Panama City was impassable. But there was a small airport in town where they might be able to get a flight over the canal. They made their way to the airport, where they met a pilot who was shuttling a party to Barranquilla, Colombia, and for $80 apiece, they could come aboard. All parties finally agreed on $50, but it was still a big fraction of their reserves.

The unexpected expense was painful, but the views of the Panama Canal from the air were spectacular. It was easy to see the man-made waterway from one end to the other, big ships moving painstakingly in both directions, and the impressive Bridge of the Americas crossing the canal near Panama City.

In Barranquilla they bought bus tickets to Cartagena, the port with a walled city and fortress where, in colonial days, Inca treasure was loaded aboard fleets of ships bound for Spain. King Ferdinand's order had been, "Get gold, humanely if you can; but at all hazards, get gold!"

They walked along the walls of the fortress in Cartagena and spent another afternoon at its grand beach, then bought more bus tickets to Quito, Ecuador. Buses were proving to be cheap and more predictable than hitchhiking, save for mudslides.

They got off in downtown Quito, at 9,000 feet the highest capital in the world. They walked around the well-preserved colonial sector, then hitched a ride to the *Ciudad Mitad del Mundo* (the City at the Middle of the World), where they straddled the "equator," one foot in the northern hemisphere and the other in the southern. In actuality, the painted stripe that runs through the town square to denote the "equator" is 240 meters south of the real equator. Returning to Quito, they saw several volcanoes, including the still-active Wawa Pichincha, with Quito situated on its eastern slope. The Battle of Pichincha, waged by Simón Bolívar and Antonio José de Sucre, brought the city's independence from Spain in 1822.

Another bus trip to Guayaquil on the Pacific Ocean descended 9,000 feet over 165 miles. As they approached the city, David and Kevin caught a glimpse of Mount Chimborazo, majestically towering 20,000 feet into the Andean sky. In the city, they walked along the waterfront, noting that Guayaquil was the primary center offering ferries to the Galápagos Islands, but again, they couldn't afford it.

Their next stop was Cusco, Peru, once the capital of the Inca Empire before it fell to the Spanish conquistadores. They spent an entire week marveling at the masonry skills of the Inca, walls constructed from huge chunks of rock that fit together perfectly and did not surrender, even to earthquakes. Native Quechua children led their pet llamas to town and offered to pose for photos for a few coins.

The two friends hung out around the Cathedral Basilica of the Assumption of the Virgin with the grand plaza in front. They once dined on roasted guinea pig, a staple in the area from ancient times. It would not become a staple in their diet. But it was a grand week mingling with Quechua people, hiking up the steep trails and down to the sacred Urubamba River, where they again imagined kayak courses through the roiling water.

Finally, they bought bus tickets to Lima, the capital of Peru and their final destination. It was another impressive descent from 11,000 feet to the Pacific Coast. The city was founded in 1535 by Francisco Pizarro, after he was named governor by the Spanish Crown.

The boys got off the bus downtown and walked around the busy streets in the colonial sector. But tourism was no longer on their minds. They were broke.

Now desperate, they slept in parks, and after five days without food, they resorted to stealing a pot pie from a street vendor who was unable to run them down. Hunger had caused them to abandon their principles.

David was certain they had done the right thing back in San Jose. But the joy of their venture had vanished. Plato's discourses on wisdom, courage, justice, and temperance were forgotten, stuffed into the bottom of his backpack. They needed help.

Kevin made a collect call to his grandmother, who had been worried about him and was pleased she could help by wiring funds for them to buy airline tickets to Miami.

During the flight, they relived parts of their journey and the tough lessons learned. They had made it all the way to Lima on their own, and the trove of memories they carried with them would remain forever. The feeling they were "the two most blessed guys in the world" had returned, thanks in part to Kevin's dear grandma.

They arrived unshaven in Miami in December. They wore lightweight clothing, sandals without socks, and their backpacks and sleeping bags were worn and tattered. The South American summer was now a North American winter. As they hitchhiked north to Akron, the wind blew cold, and there was snow on the ground.

But finally two tired, malnourished teenagers were welcomed back home by their families. David slept for almost two days.

He accepted his stepfather's insistent offer of a job at Firestone as a tank cleaner on the night shift.

"Two guys would get into these large chemical vats and scrape a silicone-like material from the inside walls of the big tanks," David explained.

The smell of the chemicals and rubber immediately sickened him, but he persevered for three days. Two days later, back at his apartment, he was awakened by his mother. She had been informed by his supervisor at Firestone that he had not shown up for work for a second day. Still feeling confused and out of sorts, he realized that an entire day had gone missing. On coming to check on her son, his mother found him jaundiced and took him to a hospital, where he remained for a week. Doctors concluded his illness was hepatitis A, worsened by the chemical exposure at Firestone. It took six months before he felt well again. The Lima street vendor had taken his full revenge.

Akron to Prescott

1975

At the age of twenty, for the first time in his life, David worked full time and had some money in the bank. His job led him to do a business exploration of the southwestern United States, where he was reminded of the majestic Andes Mountains and deserts of South America. He settled in a small town tucked into the Prescott National Forest in the northern part of Arizona.

One of his new neighbors, Dave Reynolds, owned a tree-trimming business that had grown bigger than he could handle alone, and David became his assistant. Dave, whose primary job was captain with the Prescott Fire Department, was immediately impressed with David's approach to the work. The two became good friends, and Dave encouraged him to consider becoming a fireman himself. The idea was exciting to him, and he spent the entire night going over and over what it could mean for him. Dave became his coach as David formally made his application and approached a required written exam. That was followed by rigorous physical testing and a final oral exam. David got the job, and the next thing he knew he was riding on the tailboard of a fire engine as a hose man. He excelled, and his future took a promising turn.

One class — emergency medical technician (EMT) — especially captured his interest. It was training in prehospital emergency care of sick and injured patients that included medical problems from heart attacks to obstetrics. The concept had been developed in the military services, and it was being shown everywhere that the ability of EMTs to quickly stabilize and get patients to emergency facilities saved lives.

Learning all the new material and the camaraderie with his fellow firefighters made the work feel like a true vocation. Physical conditioning was mandatory, and many of the firefighters used their down time to pursue outdoor activities. David took the idea to an extraordinary level. Almost all of his friends were athletes. Two mountaineers taught him climbing skills on the face routes and crack climbs up Granite Mountain, a landmark northwest of Prescott. Since all these activities were available within a reasonable distance, he added mountain biking, skiing, and kayaking to his list. Dave was also an avid scuba diver, and underwater exploration became another of David's interests.

He continued to add to his knowledge of emergency medicine, reading and discussing cases with the emergency room doctors who took a special interest in him.

In 1978, David saw a flyer tacked on the bulletin board at the firehouse announcing a program to train firefighters in northern Arizona to become advanced life support providers. This was a higher level of emergency medical services, involving prehospital interventions and treatments. The formal training, covered by educational leave, involved two days per week in didactic sessions in Prescott along with clinical experience working in the emergency room at Maricopa County Hospital and riding with Phoenix Fire Department rescue squads.

Nothing could have been more interesting to him than training under paramedics and emergency-care doctors in the big city. He already had considerable exposure to gruesome scenes on ambulance runs, so the shock factor wasn't a barrier to getting on with the challenge at hand. Overall, the classroom study, followed immediately by on-the-job application, was a very effective learning method for him.

In 1979, David reported to work as a certified Advanced Life Support provider with the Prescott Fire Department. He divided his time between the hours spent at the firehouse and his outdoor activities.

A few months later, David received an invitation to go on a dive trip on a sailboat in the Sea of Cortez, Mexico. The invitation was from Dorothy Centa, a former girlfriend, and her children, Holli and Shawn. Dorothy's friends, Mary and Kyle Hutchings, kept a sailboat in San Carlos, Mexico; Kyle, Mary's husband, was an avid scuba diver who needed a dive partner. The couple, along with their daughter, Carrie, who was about Holli's age, planned a trip to the small Mexican town over the Thanksgiving holiday. Mary was not a diver herself and thought it would be fun if Dorothy came along to keep her company. The two girls could get to know each other, and the kids could have fun boating, fishing, and swimming. There was plenty of room for everyone, and Kyle would be delighted if David accompanied him as a dive partner.

Sea of Cortez

1982

They worked out the logistics of the trip, and early Tuesday morning David picked up Dorothy and the kids at her place in Prescott. They set off on the long trip down to the Phoenix Valley and across the Sonoran Desert. Even in November, the biggest, hottest desert in North America looked sunburnt. Hours later, the shimmering reflection of the late afternoon sun glanced off the Sea of Cortez — a beautiful and welcomed sight.

They arrived in San Carlos about 6:30 p.m. and drove to the marina. David had not met the Hutchings, so he felt a little anxious, but he had observed that scuba divers always seemed to bear a kinship toward one another.

They parked the car and walked down to the slips. The only activity was toward the end of the pier, where three figures moved on a large trimaran named *Reina del Mar.*

The two groups greeted each other warmly. Kyle and Mary were almost finished organizing the boat. Even though it was one of the largest boats in the harbor, a tour of *La Reina* didn't take long. On either side of the main hull, which contained the wheelhouse, the galley, and the sleeping quarters, were two smaller hulls, outriggers held fast by struts extending from both sides of the main hull to stabilize the craft when the sails were hoisted. The mast extended more than thirty feet up from the main hull.

The stern was equipped with an open seating area. A narrow stairway led down to the galley and sleeping quarters. Carrie and Holli excitedly took charge of sleeping assignments. On that cool, calm night, David and Shawn chose to sleep under the stars on the decking.

The group spent the next morning, the day before Thanksgiving, getting the rest of the supplies and equipment stowed on board. David and Kyle checked the rigging, then donned their scuba gear and applied themselves to the tedious job of scraping barnacles off the hull. By mid-afternoon, everything on the boat was carefully stowed, cleaned, and polished. The divers dragged out all of their scuba gear, which took up the entire space in the stern, and rechecked wetsuits, compressed-air tanks, diving masks, pneumatic spearguns and buoyancy compensator devices (BCDs) required for deeper dives to come the next day. David's BCD, borrowed from Dave Reynolds, was a simple horse-collar design inflated by opening a push-button valve on the right side and blowing air directly into the bladder. To deflate, the same valve was held over the head, the button pushed, and the air bubbled out into the seawater. It all seemed so simple. *(See Appendix A: Buoyancy Control Devices.)*

Thanksgiving Day

6:30 a.m.

On Thanksgiving Day, David and Shawn woke up with the first rays of sunshine. There was barely any wind. They made coffee and relaxed in the stern, their quiet conversation echoing off the water. Soon Kyle, Mary, and Dorothy joined them, followed by the girls, not yet fully awake.

After breakfast, Kyle started up the inboard engine, and they eased out from the docking area. It was a small motor for a relatively big boat, but they headed out to sea, the sails remaining furled. The only ripples in the glassy water were from their own wake. It would take a long time before San Carlos was out of sight.

Good catch! Shawn Centa aboard *La Reina del Mar.* *(Photo from David Scalia's scrapbook)*

A contingent of gulls followed them out of the harbor, darting in and out while they performed aerial maneuvers to retrieve the breadcrumbs the kids tossed over the side. After the breadcrumbs were dispersed, Kyle invited the kids to tour the cockpit, where he pointed out the features necessary for navigation. Then he turned the controls over to Carrie, who showed Holli and Shawn how to read the instruments and guide the boat. Kyle periodically checked the heading as the youngsters happily occupied themselves in the cockpit for the rest of the morning.

1:00 p.m.

Kyle shut down the engine and dropped the anchor near a small rocky island, carpeted by scraggly, gray-green foliage, not a tree in sight. The kids, already in their swimsuits, got the OK to jump in; they laughed and bobbed in the water, while Kyle and David took out the scuba gear and wriggled into their wetsuits.

David Scalia before the dive *(Photo from David Scalia's scrapbook)*

The depth gauge read 110 feet. The navy dive chart indicated they could safely stay at that depth for twenty minutes. David positioned his BCD around his neck and buckled it in place. Kyle's more complicated device fit like a backpack with shoulder and crotch straps to hold it in place. They assisted each other to strap and buckle the air tanks into position on their backs. After they buckled their weight belts, the two seated themselves on the transom, feet on the seats, and slipped on their fins.

The ladies watched as the divers walked onto the decking and positioned their masks and regulators. The kids playfully swam in their path, giggling. But they soon deferred to the two giant figures hovering overhead with menacing spearguns in hand and dog-paddled off to the side.

1:10 p.m.

The divers splashed in, feet first. They waved, submerged, and disappeared below the surface. Underwater, David looked at Kyle. They gave each other a thumbs down, the diver's signal to descend.

David watched Kyle's shadowy figure glide downward with a few graceful scissor kicks, air bubbles trickling upward as he exhaled. Bright sunlight shimmered down through the seawater from above. The men slowed periodically to accommodate themselves to the increasing pressure. As they approached the ocean floor, they saw a vast field of big angulated rocks, like those on the island, protruding from the sand. For a moment, they stood on the sandy bottom, actors on an underwater stage.

* * *

People have been diving for millennia to spearfish and collect mussels, sponges, or shells. Pearl diving is an ancient, but fading, art still practiced in some parts of the world. Breath-hold divers reach depths of forty meters, using only fins and face masks. But the time underwater is limited to three or four minutes. Wars and salvage operations impelled people to probe the deep using diving bells and diving suits with fresh air pumped through a hose from the surface. (Homer wrote about military divers during the Trojan War sinking vessels by boring holes through their hulls, and Alexander the Great is said to have descended to the seabed in a glass barrel.)

Inspired by others such as Yves Le Prieur, Jacques-Yves Cousteau and Emile Gagnan invented the Aqua-Lung. In the summer of 1943 a small group of friends waded into the French Mediterranean Sea to test their "autonomous diving gear."

In 1948, the Aqua-Lung went on sale in the United States. The self-contained underwater breathing apparatus (scuba) opened up a new world for exploration by millions of passionate divers who would make their own evanescent footprints on the ocean sands.

* * *

David and Kyle searched the crevices between the undersea rocks for grouper that lay in wait for their unsuspecting prey. These large, grumpy-looking creatures were a favorite target of spearfishermen in the area. David blew some air into his BCD so he could easily drift a few feet above the sandy floor, and both divers carefully armed their spearguns. Seaweed waved lazily with the ocean current as the divers moved along. Schools of small fish ignored them, hurrying along on their never-ending search for food. Side by side, Kyle and David combed a wide field of the seafloor and noticed a few small fish weaving their way through the rocks. No groupers. However, they weren't disappointed; the sensation of being there and gliding along in the ocean current was exhilarating in itself.

The allotted twenty minutes passed quickly, and it seemed only moments before Kyle pointed to his watch and then to the surface. Divers were always on the clock. Even the best equipment was able to provide a set amount of time underwater. For a while, at least, their search had to be abandoned. There was plenty of time for a second or third dive later that afternoon. They disarmed the spearguns and prepared to resurface.

As he began his ascent, David raised the BCD valve overhead and pushed the button to release air. He didn't notice that air bubbles were not being released. Suddenly, even without kicking his feet, he realized he was ascending very quickly. Alert but still not worried, he held the valve where he could see it and pushed. There was no release of air, no bubbles. Sensing his partner was in trouble, Kyle lunged up and grabbed one of David's fins. But he couldn't hold on, and David shot straight up, out of reach.

In the next few seconds, David tried the release valve again and again. There was no response. He was in an uncontrolled ascent. His mind raced, searching for a solution. He jerked at the tube again, trying unsuccessfully to tear it free. As he continued ascending, the water pressure decreased and the BCD expanded. This caused pressure around his neck and chest.

He made himself horizontal and spread out his arms and legs to create drag. The maneuver worked, and his ascent definitely slowed.

"I've got it!" he thought.

But the effect was mercurial. Any resistance he had created was almost immediately overcome. He exhaled several times in an effort to blow off the expanding volume of air in his lungs.

Now he was completely out of control as he accelerated through the last twenty feet of seawater. The entire event had taken place in a matter of seconds.

He felt a dreadful foaming sensation moving rapidly from his ankles to his legs and upward through his abdomen. He felt his heart cavitate as if it were pumping nothing but air. A severe stabbing pain started in the center of his chest and quickly spread out to both lungs. Finally, the foaming sensation rose up into his neck. *(See Appendix B: Scuba Diving Physiology.)*

As fast as he could, Kyle followed David to the surface. Sensing a tragedy, his own body was supercharged as he swam to help. Buoyed face-up by the BCD, David was still conscious. Kyle drew closer.

Still fully aware of his situation, David calmly stated, "Hey man, I just embolized. I need a decompression chamber."

Up to that point, he had known exactly what was happening. Then bubbles had traveled up the arteries to his head and began to shut off blood circulation in his brain.

He went silent.

From inches away, Kyle watched as David's eyes closed and all movement stopped.

In less than three minutes, David's world had changed from bliss to oblivion.

1:30 p.m.

After uncoupling David's weight belt and letting it drift downward, Kyle grabbed the BCD, which was keeping David face-up, and towed him to the stern of *La Reina*, where everyone on board stood near the ladder, each knowing something terrible had happened. As they reached to help him up, David didn't reach back. It was a horrifying scene. No one dared say it, but it looked like he was dead.

Kyle was alert, running on adrenalin. "Help me get him up!" he shouted.

He pushed David's body to the bottom of the ladder, where five pairs of outstretched hands grabbed for his vest or his arms to pull him up. But the higher they lifted him, the heavier he became. Just when they thought they were making progress, they lost their collective grip, and David splashed back into the water, Holli falling right along with him. She quickly scrambled back into the boat. David was dead weight, loaded down by the heavy air tank. There seemed to be no way they would be able to haul him on board.

Now Kyle's energy waned. He had no foothold to help the others as he bobbed up and down behind David, trying to keep him within their reach. The women repositioned themselves so Shawn, Mary, and Dorothy could get a secure grip, and the three of them pulled with all their strength. They hauled David's torso into the stern, but the lower half of his body held him locked in that position until eleven-year-old Shawn reached out, grabbed David between the legs, and singlehandedly hefted him into the stern.

"He was convulsing and foaming at the mouth," Holli recalled. "It was incredibly scary because David was always the helper who would take charge in a situation like this."

Moments later, the convulsions stopped. David lay lifeless on the floor. They managed to remove his air tank and uncouple the hyperinflated BCD, pitching it aside. Once the gear was out of the way, they rolled David onto his back.

Dorothy unzipped the wetsuit and saw his chest move — shallow breaths. He was alive.

Kyle pulled himself into the stern and removed his equipment. Exhausted, he staggered to the helm, started the motor, and pointed *La Reina* toward San Carlos.

David was comatose. Had he aspirated seawater into his lungs, it would likely have killed him. Had the others failed to heft him onto the deck, he would have died right there at the back of the boat. The sight of such a vigorous young man lying helpless on the floor was gut-wrenching.

The same bewildering thought struck each of his friends.

"Now what?"

1:40 p.m.

There was no wind to power the boat, so three knots per hour was the pace. Time was of the essence, and it was going to take hours to get back to shore. As Kyle clung to the wheel, he used body English to eke out everything the engine had to give. Everyone knelt near David, who drew short, rapid breaths followed by disturbing pauses with no breathing at all. He was still unconscious, and the gravity of the situation sank in.

2:05 p.m.

Holli dedicated herself to searching the horizon for help. Suddenly, she tensed up, focused.

"Look!" she pointed, "There's another boat up there!"

They all stared in the direction she pointed. There was a boat with two figures aboard, staring at them. It seemed to appear out of nowhere. The two fishermen returned the urgent waving as they waited for the trimaran to come closer. The craft was a *panga*, a simple fishing boat with a big outboard motor. Kyle and Mary knew enough Spanish to communicate with them at least a little. When they were close enough to speak, they learned the fishermen were from Guaymas, an industrial port south of San Carlos.

The pilot of the vessel maneuvered his boat up to the stern of *La Reina*. One of the fishermen peered over the edge to see David lying helpless on the floor. Further words weren't necessary. Without hesitation, they put their gear aside to clear a space for him, and they motioned to bring David on board.

This would be the first in a series of incredibly timed events, impossible to explain. Holli had spotted the boat, a much faster craft with a pair of Spanish-speaking fishermen willing to help, which again seemed to appear out of . . . nowhere. This connection at sea would save hours of critical rescue time.

2:20 p.m.

A blanket served as a litter and a bed. The fishermen helped Kyle and Shawn slide David, still in his wetsuit, onto the floor of their boat.

"*Vamos a Guaymas,*" said the Mexican pilot.

A quick decision was made: Kyle, Dorothy, and Shawn boarded the *panga,* leaving Mary and the two girls to bring the boat back to San Carlos. Mary was familiar with piloting *La Reina,* and there was plenty of fuel.

Kyle wrote the heading on a pad near the wheel. He would contact the manager at the marina in San Carlos as soon as he could.

The situation was tense. Mary would be alone in the middle of the ocean, and Kyle, David, Shawn, and Dorothy were heading, in the company of strangers, in a different direction. For all anyone knew, the fishermen could be pirates who, once out of sight, would rob the *norteamericanos* and toss them overboard.

But the plan was the only thing that made any sense. Kyle felt that he was needed in the rescue boat, but he watched hesitantly as Mary and the girls bravely waved goodbye. The Mexican pilot gunned the engine, and the rescue party of five took off for the mainland. A four- or five-hour trip would be reduced to about sixty minutes.

Guaymas, Mexico

1982

Thanksgiving Day, 3:30 p.m.

For a little more than an hour, the rescue craft, throttle wide open, hurtled over the swells. The pilot powered down the engine as they pulled into the fishery near Guaymas. One of the fishermen jumped out of the boat as it was being tied to the pier, and he took off running. He returned shortly, backing a small pickup truck close to the pier.

"*Vamos al hospital,*" he said as he anxiously motioned to bring David to the truck.

The others worked quickly and, using the blanket, carefully lifted David out of the boat and into the bed of the truck. The transfer completed, Kyle and Shawn scrunched down next to David. Dorothy rode up front with the fisherman-rescuer. They took off to shouts of "*Buena suerte!*" coming from the lone fisherman on the pier.

But the Mexican rescuers weren't finished. Their small truck became another rescue vessel as it raced away from the dock carrying David and his friends in the direction of a medical facility. The dedication of those two strangers was far beyond expectation.

3:45 p.m.

Only a few blocks past the exit gate of the fishery, Kyle and Shawn witnessed David's color change from pink to blue. Kyle began chest compressions and mouth-to-mouth resuscitation as the truck rumbled over an uneven road. Then the truck driver spotted an ambulance.

"It just showed up," Dorothy said. "It was coming one way, and we were going the other way, and whoever was driving the truck waved him down. It was like a miracle."

The two vehicles pulled over together. Kyle, a bit unsure of his compression technique and now tiring, paused and looked up as the excited truck driver explained the situation to the ambulance driver and the two attendants. No one in the truck noticed David had suddenly stopped breathing.

The two ambulance attendants immediately saw the ghostly color of David's face.

Their reaction to the impending disaster was swift. They began cardiopulmonary resuscitation (CPR) as David faced another life-threatening crisis.

The ambulance driver hurried to bring an Ambu bag and an oxygen tank to the medics working over their patient in the bed of the truck. After finding a pulse at the base of David's neck, the attendant placed a plastic mask tightly over David's nose and mouth and pumped the Ambu bag, while the second medic connected the oxygen tank with plastic tubing. David's chest moved passively up and down in response to the pumping, and within three or four minutes his normal color started to return.

Kyle helped the attendants transfer David and the equipment into the ambulance. While the first attendant kept pumping the Ambu bag, the second listened to his heart and his lungs and gave a "thumbs up."

The fisherman motioned for Kyle and Shawn to join him in the truck, and they followed close behind the ambulance. They skidded to a stop at the hospital, where their rescuer pointed to the entrance. Kyle faced him, shook his hand firmly, and said, "*Gracias,*" frustrated with his inability to express his gratitude more adequately in Spanish. Then he emptied his pockets of most of their Mexican currency and placed it into the hands of . . . an angel?

"*Gracias,*" Kyle said again, shaking his head, an expression of appreciation and disbelief at the same time. Then he turned toward the hospital, put his arm around Shawn's shoulders, and the two of them hurried toward the door.

"*Buena suerte,*" the fisherman shouted, waving his hand in the air one last time as the door closed.

A nurse directed Kyle and Shawn to a waiting area where they sat together, collecting their thoughts. They marveled to themselves how the ambulance appeared within moments after David had stopped breathing. In minutes, his heart would have stopped beating, followed by the loss of all brain activity. He would never have made it to the hospital alive without that encounter.

Although young people with traumatic injuries statistically fare a little better when treated outside a hospital, CPR is effective in only a small percentage of patients. Even with their experience in CPR, the ability of the crew to resuscitate David in the bed of a pickup truck was extraordinary.

4:00 p.m.

The ambulance backed up to the dock, and the attendants quickly carted David into one of the bays. Moving alongside the gurney, the ambulance attendant dutifully continued his rhythmic squeezing of the Ambu bag. A doctor appeared and listened carefully with his stethoscope. When he tried to rouse David, there was no response. He asked the two attending nurses to help remove the wetsuit so he could complete his examination. Finding it difficult to remove, the nurse cut it off.

As David's body was being jostled to rid him of the impeding garment, he moaned and then, from down the hallway, Shawn heard screams he would never forget — loud, almost inhuman, screams that echoed into the waiting area. They all knew it was David.

The unspoken question was, "Were these his last gasps?"

After his outburst, David fell still again, his body soaked in sweat. One of the nurses toweled him off and hooked up a heart monitor. The doctor started his examination anew. He asked the ambulance attendant to stop ventilating as he listened for lung sounds. The breaths were irregular — shallow, rapid breaths, then long pauses — Cheyne-Stokes respirations, a breathing pattern associated with impending death and probably caused by damage to the respiratory centers in the brainstem from the air bubbles.

But a resilient heart kept its pace. The Ambu bag was put aside; now oxygen flowed through nasal prongs. A chest X-ray was taken at the bedside, and blood tests were sent off.

The group stood as the doctor entered the waiting room. He introduced himself and motioned for everyone to sit. Then he sat, too. He was a thin man with a few gray hairs visible. He carefully scrutinized the three Americans and addressed them in understandable English. As he spoke, he looked from one to the other. He informed the group that David remained unconscious except for the short outburst. He was receiving oxygen, an IV was running, and his blood pressure, low on arrival, had improved.

"This is a serious situation," he said gravely. "The next twenty-four hours are critical."

Kyle broke in to relate that David was a paramedic, and he had told him he needed a decompression chamber before he passed out. In a matter-of-fact tone, the doctor informed the group there was no chamber available anywhere in this part of Mexico.

They knew this hospital wasn't equipped to deal with his injury. A decompression chamber was the only equipment that offered any chance of help, and there was none in this part of Mexico.

Dorothy remembered a tour David had given her of the communications center at the Prescott Fire Department (PFD). He had pointed out that PFD was part of an extensive network that connected with emergency facilities all over the United States and beyond. Her first step would be to call the Prescott Fire Department.

When the operator answered, "Prescott Fire Department," Dorothy asked to speak to Captain Dave Reynolds. Kyle listened over her shoulder and reminded her to be sure to tell him about the decompression chamber. Dave assured her he understood completely and said he would get back to her within thirty minutes. A series of protocols were immediately set into motion. The fact that Dave happened to be on duty at that moment is another link in the series of events that saved David's life.

Then Kyle made a telephone call to the manager of the marina in San Carlos, who promised to stay on the lookout for *Reina del Mar* at the San Carlos marina.

5:15 p.m.

Dave initiated multiple calls seeking assistance. A decompression chamber was located aboard a ship, USS *Dixon*, at the San Diego naval base. US Navy Captain M.A. Heuberger granted permission for use of the government equipment, if necessary. Air Evac in Phoenix directed PFD to contact Critical Air Medicine in San Diego for an evacuation flight from Mexico. The San Diego dispatcher told Dave they could get to Guaymas in three hours but cautioned that Mexican immigration officials had often held up evacuation flights before clearing them to return to the United States.

"Send the plane," Dave said. "We'll take care of the rest." He hung up the phone, asking himself, "But how?"

5:30 p.m.

Caborca!

Dave had been a member of the committee that had made Prescott and Caborca, a town north of Guaymas, Mexico, "sister cities" in 1972. Caborca's mayor had visited Prescott, and Dave had been there to greet him. The "sister cities" idea was a gesture of goodwill that resulted in Prescott's donation of decommissioned fire engines and ambulances to the Mexican town with a good deal of fanfare. Dave sorted through a stack of business cards and found two from Caborca officials. Using the phone number from one of the cards, he got an international operator to call Caborca's city hall, and the city manager made it clear the promise of goodwill was in earnest. He assured Dave he would contact Mexican immigration officials at the airport in Guaymas. The prospect of a disastrous holdup over immigration clearance had, in all likelihood, disappeared. Precious minutes, perhaps hours, were saved.

"Dorothy was panicky when she called," Dave said, "but after I got the information I needed, we set up the evacuation. I called my contacts in Caborca, and they helped get permission for the flight to land in Mexico. The pilot and copilot had to stay with the plane, but the doctor and nurse would be able to get to the hospital without any delay. No formal permits would be required."

Dave finally called back from Prescott to tell Dorothy a rescue plane was on its way to the airport in Guaymas. Just before 8:00 p.m., the plane carrying Dr. Paul Phillips and Nurse John Carter landed in Mexico. An immigration official, obviously following orders from higher-ups, met the plane as soon as it taxied to a stop and allowed the medical personnel to deplane, directing them to an ambulance waiting nearby. The flight crew watched as the ambulance sped away and remained in the plane while it was refueled.

One hour later, the ambulance returned with David and Dorothy on board. The official quickly checked Dorothy's documents, while the crew settled David into the plane. They took off for San Diego, where a helicopter would be waiting to transfer the patient from the airport to USS *Dixon.*

When the helicopter touched down at the helipad at Naval Base San Diego, Dr. Phillips reported to Lieutenant Commander Greg Adkisson that the patient "may be DOA. We may have gotten here too late."

San Diego

1982

Thanksgiving Day, 11:45 p.m.

"He's not dead until he's in our chamber and dead," the lieutenant commander replied. Far from being hyperbole, the statement was based on years of experience with decompression treatments. Dr. Greg Adkisson had first heard the mantra from Corpsman Jim O'Connor, a trusted medical assistant who presently served as the inside attendant in the chamber aboard *Dixon.*

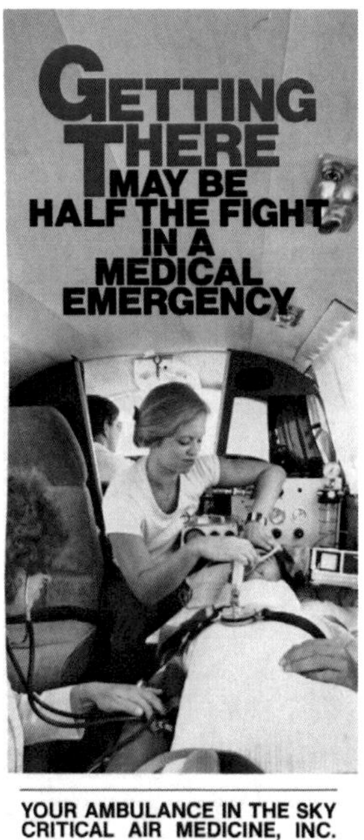

Critical Care Medicine brochure *(Photo from David Scalia's scrapbook)*

"Cocky as it was," Dr. Adkisson said, "it was a great line, and we always lived by it."

It was obvious the chances of survival in this case were poor, and the doctors had to consider the possibility that even if David survived the treatment, there was a serious risk he could be an invalid for the rest of his life. At this point, there was no way to predict the outcome. They were all scuba divers, and each was about David's age; not one of them had any doubt about what they would want if they were in David's position.

Dr. Phillips continued his case report as he, Dr. Adkisson and Corpsman O'Connor hustled ahead of the corpsmen who deftly hauled the stretcher up a series of stairs, across a gangplank, and onto the quarterdeck where the medical facilities were located.

The helicopter immediately readied for takeoff to return to UCSD Medical Center with Dorothy and

The Life Flight helicopter that transported David, Dorothy, and Dr. Phillips to USS *Dixon* from San Diego International Airport *(Photo from David Scalia's scrapbook)*

Nurse Carter aboard. She watched from her seat on the copter as it lifted off. She saw the sailors carry David over the pier to a steep stairway, where they hefted the stretcher up the stairs to the medical deck. She lost sight of him as the helicopter turned toward downtown San Diego.

Day 2, 12:15 a.m.

"Back-to-back sessions of Table Six at six atmospheres?" Dr. Phillips asked as Dr. Adkisson explained his plan. *(See Appendix C: Decompression Chambers and Hyperbaric Medicine.)* Dr. Phillips knew the medical literature and stuck to what was dictated for patient care. True, the US Navy protocols for decompression therapy were still evolving, but nowhere did the Bennett textbook suggest doubling up the sessions nor increasing the pressure setting.

"I remember thinking at the time," Dr. Phillips said, "'Why are we doing this?' There surely was no guidance to do that in the literature." But he deferred to Dr. Adkisson's considerable experience with hyperbaric medicine.

David was carried into a large room containing a cylindrical chamber that looked like a small submarine with an entry door at the rear and an array of gauges and controls on the outside. A large window on the side allowed the attendants outside to visually communicate with those inside.

Once David was inside the chamber, Corpsman O'Connor hooked up the cardiac monitor. No rhythm was detected.

Aft view of USS *Dixon* with stairway to medical deck *(Photo from David Scalia's scrapbook)*

"No pulse, I can't feel a pulse!" Dr. Adkisson shouted and started chest compressions.

The sheet covering David was tossed to the floor. He lay naked except for the wires pasted to his chest. Dr. Phillips connected the respirator inside the chamber. Three or four minutes later, Dr. Adkisson stopped compressions as Corpsman O'Connor compressed the femoral artery.

"We have a pulse!" he shouted. The cardiac monitor now displayed an irregular rhythm, but the heart had resumed its pumping action.

The decompression chamber on board USS *Dixon* *(Photo from David Scalia's scrapbook)*

Dr. Adkisson was faced with David's third cardiorespiratory arrest. And almost incredibly, another course of CPR was performed successfully. The doctor knew the only chance for survival was aggressive treatment in the hyperbaric chamber. He and Corpsman O'Connor would remain inside the chamber throughout the experimental protocol, while an outside attendant monitored the oxygen and pressure settings. Once again, the timing of a critical setback made a successful intervention possible. Even though David's survival at this point was unlikely, this rescue team would not fail. David was still alive.

Dr. Adkisson saw that Dr. Phillips could barely stand for lack of sleep. "That's why we have a cot on the other side of the room. Out!" he said. "See you in a few hours."

The chamber was sealed and pressurization began.

Dr. Adkisson and the corpsman stayed inside as the chamber door was sealed shut. No one would be able to enter or leave for the next nine hours. *(See Appendix C: Decompression Chambers and Hyperbaric Medicine.)* The pressure rose, and 100% oxygen flowed from the respirator into David's lungs. It was eleven hours since the accident. David had already been resuscitated three times — in the back of a pickup truck, on the plane, and inside the decompression chamber.

"Until we get him recompressed in the chamber," Dr. Adkisson said, "you never know what can happen. It's kind of like hypothermia. You rewarm and give him that chance. Every once in a while, you compress some critical bubble [that blocked an artery], and something amazing happens."

As the pressure increased, they saw the EKG change to a regular rhythm, and the pulse strengthened. But nine hours later, David was still unconscious.

Dr. Phillips rode with David in the ambulance to UCSD Medical Center, where he was re-evaluated in the emergency department and admitted to the ICU. Dr. Tom Neuman, staff doctor in charge, reviewed the case with Dr. Phillips. The "new admit" remained unresponsive and on supportive care.

Like Dr. Adkisson, Dr. Neuman had extensive experience in undersea medicine courtesy of the US Navy. The two discussed the situation by telephone and agreed they should continue with shorter but daily treatments in the decompression chamber and see what happened. "But," Dr. Neuman said, "All of us were relative 'newbies.' We were making decisions based upon what made physiological sense and what we felt was best medical practice. There really was no road map for us to follow at that time."

Back in Prescott, the news of the tragic event had spread quickly. Everyone in the police and fire departments knew about the accident.

Word went out to the press, and in the unique way a small town can, Prescott rallied for one of its first responders. Helpless to do anything directly, many people prayed as they waited for updates from San Diego.

Day 3, 9:00 a.m.
Second Hyperbaric Treatment

Dr. Phillips continued to follow his patient closely, and he again boarded the ambulance as it shuttled David back to USS *Dixon* for the second time. David was still intubated, and his condition hadn't changed. Efforts by doctors to rouse him earlier that morning were unrewarded.

The second hyperbaric session lasted a little more than two hours. Dorothy was waiting in David's room when he arrived back at UCSD Medical Center. As she sat next to his bed, she saw one of his eyelids twitch.

She pushed the call button for the nurse. Excited, the two of them stared intently for some time, but there was not even the tiniest further movement. The excitement turned to disappointment. The nurse gave her a pat on the arm and left the room.

Dr. Neuman arrived with a small group of medical residents. They opened David's eyelids and peered in with lights, spoke to him in loud voices, and shook him. There was still no response. The doctor looked serious when he turned to Dorothy, informing her the situation hadn't changed. It wasn't necessarily bad news, but he told her everyone had to be patient.

Day 4, 7:00 a.m.
Third Hyperbaric Treatment

On the Sunday after Thanksgiving, another two-hour session was completed. The doctors still saw no noticeable change in his condition.

Day 4, 6:00 p.m.

Before he left the hospital that day, Dr. Neuman stopped by one more time to review the chart and do another brief examination. He gently moved David's head from side to side.

"Wake up, David," he said in a loud voice. Now he thought he saw a facial reaction, but he couldn't get David to open his eyes or squeeze his fingers. When he moved David's hands and arms, the muscle tone felt more lifelike. He checked the nursing notes and learned that for a brief time in the afternoon David had seemed to react to voices. Dr. Neuman left the hospital more optimistic, even a bit excited. Could this patient survive after all he had been subjected to? His extensive experience had taught him not to expect miracles, but when the patient is given a chance, unexpected outcomes do happen.

Dr. Neuman was the senior member of David's medical triumvirate. A super-achiever, he completed training in various naval schools and served as a naval submarine medical officer, naval undersea officer, and naval saturation qualified medical officer. *(See Appendix D: Saturation Diving.)*

Dr. Neuman's navy experiences brought him to San Diego and UCSD, where he eventually acquired a hyperbaric chamber and became director of the Hyperbaric Medicine Center at UCSD Medical Center. (*See Appendix E: San Diego Reflections.)*

Day 5, 7:00 a.m.

The transfer from UCSD Medical Center to USS *Dixon* was now a routine. The ambulance arrived at the hospital at 7:00 a.m. Dr. Phillips would be ready and waiting to join them. When they arrived at the pier, two seamen would unload the stretcher and wheel it to the bottom of the stairs, where they would meet "the brigade" that passed him up the stairway, doing their best to keep David in a horizontal position. Corpsman O'Connor spent his early mornings readying everything inside the chamber. Dr. Adkisson would do a rapid assessment outside the chamber, review the lab tests, and double-check the protocol for the day with the attendants. Besides Corpsman O'Connor, there was a team of specialist corpsmen assigned to the hyperbaric chamber. The sailors would then maneuver David, with a portable ventilator between his legs, through the chamber door and settle their patient onto the bed inside. Corpsman O'Connor was always with David in the decompression chamber, keeping vigil beside him. Occasionally, Dr. Adkisson joined them inside the chamber.

Today, Dr. Phillips continued to hover outside, having resigned himself to the likelihood that "by the third day, the patient would be in a vegetative state and would never come back."

"After seventy-two hours of coma," Dr. Neuman said, "the likelihood of neurologically intact survival approaches zero. The neurologists wanted to contact the family to suggest organ donation. I had to intervene and explain in cases of arterial gas embolization that was not the case, and it was not yet time to give up hope."

Corpsman O'Connor always greeted David as they started the treatment, a brief one-sided chat that never elicited a response. There was no difference this morning as he secured the door, checked the oxygen flow, stationed himself, and watched the EKG monitor for a few minutes, the electronic blips marching orderly across the screen, confirming David was still alive. Corpsman Jim and Dr. Adkisson anticipated the adjustments of their own bodies to the steady increase of the pressure inside the chamber because, along with David, they were being treated every day, minus the 100% oxygen.

Twenty minutes into the treatment, while Dr. Adkisson pored over a medical journal, something rustled beneath the sheet covering David's body. Corpsman O'Connor leaned in closer to make sure he had heard something. David's head began to move side to side on the pillow, dragging the respirator tubing with him. He began moving his arms and legs like a newborn baby. Then his eyelids opened.

Corpsman O'Connor looked in astonishment at Dr. Adkisson, who had dropped his journal to the floor as he stood to lean over his patient. Incredulous and joyful at the same time, he peered into hazel eyes looking back at him. David was returning from wherever he had been for the past five days.

Using the navy phraseology for decreasing atmospheric pressure, Dr. Adkisson called over the intercom, "Bring us back up."

Dr. Phillips peered through the observation window. As he waited for the chamber to be safely depressurized to the point where the door could be opened, he stood by the window while Dr. Adkisson watched David move agitatedly like, "What the heck is going on?"

David's eyes moved around the chamber, appearing to attempt to comprehend an incomprehensible event.

Dr. Adkisson noted, "On this fourth day, he had made a dramatic change, waking up, but not yet completely."

Dr. Phillips revisited the moment years later. "I remembered pooh-poohing the idea of using untested criteria, not proven to be effective," he said. "Then when I saw him recover, I thought, 'This is really instructive.' I mean these guys were right to try so hard. We broke all the rules in terms of decompression tables. Looking back, it was a case of the best person doing his best work and hoping it would come out OK. For sure, there was no guidance to do that in the literature.

"The navy was just amazingly collaborative," he continued. "It was wonderful then, turning their facilities over for civilian use for David. It was a great lesson for me at the time. Today, in cardiology, we use cooling protocols, and when people have been cooled immediately after a cardiac arrest, they may wake up five days later and recover." *(See Appendix E: San Diego Reflections.)*

The doctors knew the decompression chamber would give their patient a chance. Many things had to come together fortuitously, but in the end it would be up to David to find the will to live. He did.

Dr. Adkisson put two fingers into the palms of each of David's hands saying, "Squeeze if you can hear me."

It was a far cry from his usual grip, but it was a grip. Even Dr. Adkisson, with his considerable experience in diving injuries, stared for a moment in wonder at his patient, realizing this was one of those moments a medical team is uniquely privileged to enjoy — saving a life.

David continued to be agitated, and Dr. Adkisson saw he wasn't fighting the medical people, he fought the endotracheal tube. He disconnected the respirator and watched as David breathed on his own.

"I extubated him," Dr. Adkisson recalled. "I made the decision to extubate him in the chamber. It was kind of a youthful, I won't say indiscretion, but an arrogant maneuver. Who knows what could have happened."

As it turned out, the first interpretable interaction between doctor and patient was when David closed his eyes in relief after the tube was gone.

Before Dr. Adkisson made his notes, he made a joyful phone call to Dr. Neuman to tell him he was in for a big surprise. It was ninety-two hours, almost four full days after the accident.

"Like many things in medicine, this was a risk/benefit consideration," Dr. Adkisson said. "We had used this treatment [back-to-back sessions] before for serious cases with some success, and we felt it would give David the best chance for recovery. Too deep, too long at increased levels of oxygen have the potential to lead to neurological symptoms, up to and including seizures. It is the reason that air breaks are put in between periods of breathing oxygen in all navy treatment tables. [But] that was partly why I would 'ride the chamber' that day."

Even if David had developed a seizure, Dr. Adkisson was confident the airway was protected.

Day 5, 1:30 p.m.

Dr. Neuman waited as David was wheeled back into his room in the ICU. Dr. Phillips followed, eager to see Dr. Neuman's reaction. David tried to orient himself to his surroundings. He moved his head and his eyes as if he were taking everything in. Then his eyes focused directly on the two doctors. Their biggest fear — that David would have irreversible brain damage — was over. It was obvious he could see, hear, think, and move all of his extremities.

Dr. Neuman's pace quickened as he walked down the hall to meet with the residents. He had been reminded that he had chosen exactly the perfect profession for himself. "The physician teams at UCSD Medical Center and on *Dixon* were vitally important in achieving this remarkable outcome," he reflected, "but there is just no end to the problems that occur in a comatose, ventilated patient on a day-by-day basis. Care delivery by nurses, technicians, and therapists often nip these problems in the bud . . . long before they become awful complications that can ultimately undo all of the advances a patient may have made."

Day 5, 4:30 p.m.

After the awakening and the removal of the breathing tube earlier that day, David was in a lot of pain. The doctors probed around for its source. It proved to be in the upper chest area, not at all uncommon in people who have had CPR. Broken ribs were expected. But because it was localized to the area where the cuff of the endotracheal tube was located, it was also possibly due to trauma caused by the cuff, which probably would have expanded to several times its normal size before it ruptured during the air evacuation from Mexico. A chest X-ray revealed recent rib fractures, but it was reassuring to find the heart size was normal, the lung fields were well aerated, and there was no pneumothorax. The decision was to give him pain medications and continue observation.

David was awake but not yet able to utter anything comprehensible. All he could produce were scratchy, undecipherable sounds. He searched for words he couldn't find. He moved his arms and legs almost constantly as if trying to reestablish control — as if he were inside a body that belonged to someone else. The three doctors discussed the situation, concluding that David tolerated decompression well and it was possible they could see further healing with a few more treatments using the standard US Navy protocol.

Day 6, 7:00 a.m.

The next trip to USS *Dixon* was for the sixth treatment. It was a good day. Now free of the trach tube, David responded to Corpsman Jim O'Connor's greeting with a scratchy, but understandable, "Hello" of his own. The corpsman had a hundred questions in mind for his patient. But from the start, David's responses weren't making much sense. He would start a sentence, hesitate, shake his head, and smile. Corpsman O'Connor stopped probing.

Before David left in the ambulance, Dr. Adkisson announced that the next day would be the final treatment. David nodded his understanding.

Day 6, Noon

David was transferred out of the ICU and into a regular room. It had been five days since he had anything to eat. He started with clear liquids, a precaution because of the chest pain. He swallowed gratefully with no trouble.

After noon, the team of doctors came by. One of the residents gave him a small tablet and pen and asked him to try writing his name. Obtaining the signature was a serious part of the medical evaluation. When compared to the one on his driver's license, it confirmed his identity and helped to assess his mental and physical progress. Pen in hand, David hesitated over the tablet, then scratched a few unorganized lines on the paper. Frustrated, he shook his head and put the pen down. The doctor told him he needed to practice; it would take time. David spent much of the rest of the day trying to accomplish the simple task without success. And he was exasperated every time he tried to speak. He still couldn't find the right words.

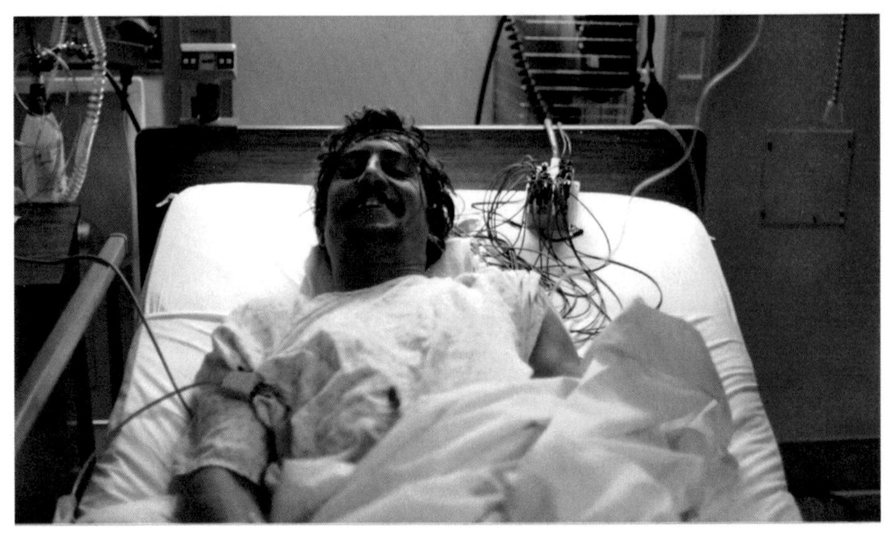

David at UCSD after transfer from ICU *(Photo from David Scalia's scrapbook)*

Day 6, 4:00 p.m.

His right ear hurt, and his teeth ached relentlessly. One of the residents found that David's right eardrum was perforated, the remnants of the delicate membrane barely visible. It probably was collateral damage from the decompression treatments, but it may also have happened during the dive itself or on the airplane flight. An attending doctor explained to the residents that pressure changes can cause serious damage to tooth fillings, inlays, and sinuses because the high pressure can result in air pockets trapped in spaces with no open connection to the outside, a setup for infections. *(See Appendix F: Barotrauma to the Head [Squeezes].)*

David was wheeled to physical therapy, where he made some progress. To begin with, they helped him with passive exercises, and before he left, they helped him to stand. He stood for about a minute before the attendants had to stabilize him. But walking was still out of the question.

Back in his room, David continued working with pen and paper, and while he still couldn't write his name, the scribbles were looking more organized. Perhaps it was because he made progress, but with each small achievement he made, he became aware of a new problem he hadn't recognized before. For example, when he focused on doing simple calculations in his head, like addition and subtraction, the answers wouldn't come. He had always been good at math, but as hard as he tried, he couldn't put two and two together.

Day 7, 7:00 a.m.

It was another good day. David expressed himself better. On the way to USS *Dixon*, David and the ambulance attendants worked out a surprise for the US Navy. They would perform their usual routine of backing the ambulance up to the pier, but when they opened the rear door, the attendants would help David onto his feet and he would stand — by himself. They all thought it a splendid idea and could hardly wait to see the reaction from the sailors.

David was still a little shaky, but he managed to slip out onto the dock as planned and stood there, a big smile on his face. As anticipated, the sailors were astounded. The two managing the stretcher gasped, and the astonished expressions of each member of the "litter brigade" quickly became delighted laughter and then whoops as the rest of the crew hurried to the rails on several decks overhead, throwing their caps into the air in celebration of the role they had played in this rescue.

"I remember the salute that last day," Dr. Adkisson said. "The people were all aware. The guys had talked and got as interested in the case as we were. [David] was essentially dead, and we turned it around. We got a pulse back. People followed the case, wondering how everything was going. We treated a lot of diving injuries; they would come in, get treated, and be done. The fact that he walked away — they liked that. This was certainly one of if not the most serious case we had ever treated, and he walked away from it."

Standing was the best David could do at this point; he could do nothing but stand, and he wasn't going to be able to do that much longer. Lifting his arms over their shoulders, two of the sailors walked him over to the infamous stairway, then carried him up as if he weighed nothing at all, while the others continued to cheer.

David had spent twenty-two hours inside the pressurized steel tube on *Dixon*. When the last treatment was over, they put him into a wheelchair and went out on deck, where a celebration began, complete with cupcakes and punch, compliments of the US Navy. Over the next thirty minutes he would

USS DIXON
Dependents Cruise '82

Submarine Tender USS Dixon AS-37
presents to
David M. Scalia
HONORARY CREW MEMBERSHIP

Effective date: *20 November 1982*
Location: *Dixon Pier, Sub Base, S.D., Ca.*

N. A. HEUBERGER
COMMANDING OFFICER

Certificate of Honorary Crew Membership, USS *Dixon*, signed by Captain Heuberger *(Photo from David Scalia's scrapbook)*

say, "Thanks" a hundred times. Captain Heuberger came below and presented David with a signed certificate that read, "Submarine Tender USS *Dixon* AS-37 presents to David M. Scalia Honorary Crew Membership."

Before leaving, David motioned for Corpsman O'Connor, his constant companion in the chamber, to come closer.

"Can't thank you enough, man," David said.

It was the first complete sentence he had uttered since the accident.

It was the last time David would see Dr. Adkisson, so he had a question that needed an answer.

Words began to flow. "Can I ever go scuba diving again?" David asked.

Dr. Adkisson thought for a moment, deciding he had to err on the safe side. He stressed caution. But he was convinced that, once a patient had fully recovered, there was no more risk of decompression syndrome than for other divers. "But," Dr. Adkisson emphasized, "you have to be absolutely back to normal."

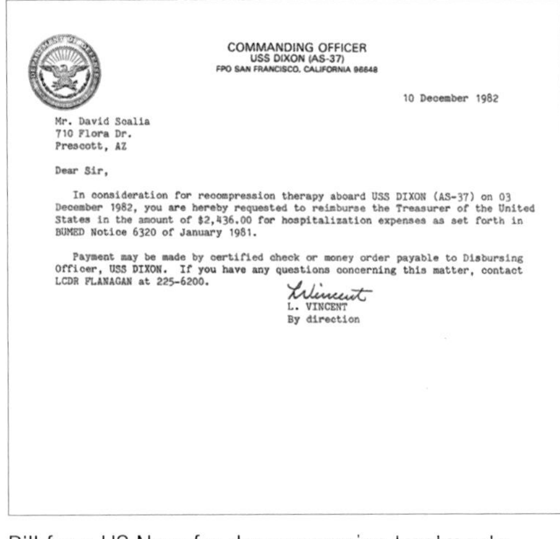

COMMANDING OFFICER
USS DIXON (AS-37)
FPO SAN FRANCISCO, CALIFORNIA 96648

10 December 1982

Mr. David Scalia
710 Flora Dr.
Prescott, AZ

Dear Sir,

In consideration for recompression therapy aboard USS DIXON (AS-37) on 03 December 1982, you are hereby requested to reimburse the Treasurer of the United States in the amount of $2,436.00 for hospitalization expenses as set forth in BUMED Notice 6320 of January 1981.

Payment may be made by certified check or money order payable to Disbursing Officer, USS DIXON. If you have any questions concerning this matter, contact LCDR FLANAGAN at 225-6200.

L. Vincent
L. VINCENT
By direction

Bill from US Navy for decompression treatments
(Photo courtesy of David Scalia)

David thanked his doctor for the advice, for all he had done, and then bid him goodbye. It was a day he would never forget — the people and USS *Dixon*, the place where his miraculous recovery had happened. *(See Appendix E: San Diego Reflections.)* The sailors helped him back down the infamous stairway and out to the ambulance for the last time.

Now that the decompression treatments were over, an ear specialist did minor surgery repair on the eardrum and placed a myringotomy tube across the repair to prevent vibration and allow the membrane to heal. David was warned not to get any water into that ear until the eardrum was completely closed. *(See Appendix F: Barotrauma to the Head [Squeezes].)*

His toothaches gradually subsided. The doctors attributed them to pressure changes during the dive as well as the decompressions. He would need to see a dentist back in Prescott for X-rays and likely need to have some of his fillings redone.

Day 7, 2:00 p.m.

David continued to get stronger. In physical therapy, he walked the length of parallel bars several times, stabilizing himself with his arms.

Dr. Neuman, already recognized as one of the leaders in the field of diving medicine, announced, "In my experience with treating diving accidents so far, I've never seen anyone come back from so severe an injury, much less recover fully." *(See Appendix E: San Diego Reflections.)* Then, he contradicted Dr. Adkisson and warned David *he should never go scuba diving again.* He didn't want David to place himself at risk.

Feeling that he would be released from the hospital soon, David yearned for that time to come.

Day 8, 11:00 a.m.

David now understood what others said, and there was noticeable progress with his speech. It wasn't perfect, but he could get his point across if the listener were patient.

As the days passed, his confinement in the hospital became increasingly antagonistic to his independent nature. His desire to get home and back to his own life intensified. He became fixated on the unrealistic idea that he had to get back to work immediately. When Dr. Neuman came by on his morning rounds, David became petulant.

"I need to get out of here," he said.

"Dave," Dr. Neuman said, "you aren't ready for discharge."

"Can you fix the problems in my brain?" David asked.

"Only time can do that, but I am sure you will continue to get better."

"Then I want to get better at home."

"You're not strong enough yet," the doctor said. "You've been through a tremendous ordeal. You have to be patient for just a while longer. Focus on walking."

Thinking he had made his point, the doctor patted David on the shoulder and left the room.

David's mood for the rest of the day was gloomy as he considered what to do next.

Day 9, 7:00 a.m.

David informed the head nurse he was leaving, an announcement that resulted in a call to the chief resident, who in turn called Dr. Neuman. In the end, David signed a form and left against medical advice, only to return shortly thereafter, feeling weak and lightheaded. The chief resident checked him back into the hospital and wrote some new orders. Knowing what had transpired earlier that morning, Dr. Neuman came by about 10:30 a.m.

For some time he stared at David, the nonverbal communication expressing the doctor's "I told you so" message. When he finally spoke, he told David he understood and was sure that only a few more days in the hospital would make the difference.

David had to get stronger. He dedicated himself to physical therapy and cooperated with the staff.

Day 12, 9:00 a.m.

Over the next few days, David remembered fragments from the time he was comatose. He recalled the recurring sensation of "going up" to the brink of consciousness from a dark, ethereal place somewhere deep below. When he was "up" he heard voices, uninterpretable because they all seemed to be speaking at once. Those memories were associated with terrible physical pain that stopped only after he willed himself to return into the dark nothingness where there were no sensations and no memories.

One memory was more like a dream. He found himself hovering, like an angel, over an emergency room, where he watched the doctors and nurses at work. Their shapes were distorted. Some had elongated faces and pointed ears, but the deformities were of no concern to him. From above he watched the doctors work on him — down below. They spoke loudly, urgently. All details in the room — the patients, the beds, the workstations — seemed vivid and real. In his mind, it must have happened when he was being admitted in the emergency department the first day.

Another memory stunningly clear to him was a voice calling his name. "David, you must make a choice. You can go up, or you can go down. But you must choose."

He chose to go up; he chose to live. The memory of that moment — the moment he made his choice — has never left him.

These were the only phenomena he remembered from his four days of unconsciousness — no brilliant lights, no insights into the future, no mystical beings. At peace and separated from his body, he hovered over the emergency room, watching the doctors and nurses working on his own body. He was given the choice — and he made a conscious decision to return to life. There was never any fear or distress, just pain when he rose up from the depths, pain that drifted away when he sank back into dark nothingness.

Others who have had near-death experiences (NDEs) have reported similar memories, such as vivid out-of-body sensations, where they saw, while hovering overhead, medical professionals working below. Some describe a floating sense of freedom from pain and complete well-being.

About half of all individuals who survive NDEs say they were given the choice to live or to die. The reason for their choice to live was often a desire to continue with the people or things they had loved in their lives and the feeling that life still had gifts in store for them.

David discussed his memories with the chief resident and Dr. Phillips. The only time he had ever been in the emergency room at UCSDMC was when he was admitted, unconscious, after his first decompression treatments on USS *Dixon.* Neither of the doctors was convinced David, considering his condition at admission, could have remembered anything about the ER. But the resident doctor was interested enough to take him back to the ER by wheelchair for a tour. As they moved through the area, David studied the faces, the nurses' stations, and the exam rooms. Absolutely nothing correlated with his vivid memory.

There were many challenges to be met and overcome in David's ordeal. How his body tolerated such devastation is an enigma even for medical professionals. But it did — he did. He walked back to the life he had known.

In his mind, this experience had brought him closer to divine intervention than he had ever experienced before. Others had made choices on his behalf — choices that kept him alive. But the preeminent choice — to return to his life — was his alone.

Prescott, Arizona

December 1982

Day 13, 9:00 a.m.

It was another unforgettable day when Dr. Neuman officially dismissed David. "You were close to being pronounced dead [when we] decided to give decompression a try," Dr. Neuman said, "but [we] did not hold out much hope."

David knew he owed his life to this medical team. His work, too, was saving lives. As a paramedic, he would have the opportunity to return the favor to others.

Upon returning home to Prescott, he arrived at Dorothy's place, where Holli and Shawn ran out the door to greet him on a cold December afternoon. The hugs were in earnest, long and tight. The last time the children had seen David, the two youngsters had faced death for the first time. In their entire lives, they had never even really known anyone who had died, especially anyone as young as David. Now he stood before them, his bright smile unchanged from before, and they found it easy to block out the memory of such a dark and troublesome topic. When they entered the house, the group of four huddled together, arms around one another.

In the recent past, David's thoughts had been directed almost exclusively to his own needs. But returning home changed his perspective. After he left the doctors and nurses back in the hospital, the enormity of what he had been through hit him hard, almost overwhelming him. Inexplicably, his life had been given back to him. Standing there, in the living room, he realized he owed his life to this little circle of people. How does one repay such a debt?

He found it difficult to speak, but he managed to say, "I will never be able to thank you guys enough." And now he couldn't hold back the tears.

David was still very weak and had lost a lot of weight. He took liquid supplements to replace lost muscle mass.

Wishing to see their comrade, a number of his firefighter friends came by to cheer him up the first night he was back at his home.

Firemen Dave Scalia, right, and Dave Reynolds after Scalia's recovery.

(Photo courtesy of Prescott Daily Courier, *1982)*

Dave, the captain, hugged him. "You're a helluva guy," he said. David answered, "You, too."

When David tried to answer questions posed by his colleagues, his problem with word retrieval was exposed. He explained the lapses were especially bothersome "because before his accident, after all, he had been a genius." The explanation convinced everyone that he was going to be OK.

For the next several days, people dropped by, usually with food, and spent time with him. Every day he bundled himself up and went out for a short walk, then sat on the porch and breathed in the wintry air until he got cold. As time passed, the visits dissipated. The hours of solitude began to wear on him. Deep down, he worried he would never be strong enough to go back to work.

> Dear Friends;
> Words cannot express the gratitude in my heart. Your compassion and prayers did not go unnoticed as God has let my life go on. I am presently doing well and should be back ridding the Rescue Truck by the time you recieve this card.
> It is my whole hearted desire that I be given ample opportunity in life to display the same Christ-like compassion you people gave to me.
>
> God Be With You
> David Scalia

David's handwritten note to his friends *(Photo from David Scalia's scrapbook)*

Christmas decorations were all around. He had lengthened his walks considerably, and the sparkling lights brightened his spirits. He wrote thank-you notes to his friends who had supported him during his ordeal. He longed for the camaraderie at the firehouse, and he occasionally stopped by the station, where short conversations with the other firefighters heartened him, at least for a while. Finally, he asked Dave about coming back to work, who reminded him that firefighters were a team, and everyone had to be able to contribute 100%. And David couldn't even think about coming back to work without medical clearance.

With each week that passed, he felt stronger and more confident. Physically, he knew he was short of 100%, but could he pass a medical exam?

He wanted his life back again. Would a doctor be able to detect anything that might bar him from his profession? He could not allow that to happen. Concerned his recent medical history could have a negative influence on the doctor's conclusions, he decided to gloss over it. This would ensure a final opinion based only on the observation of how he was now, leaving out any bias introduced by his past history. He couldn't wait any longer; he made an appointment with a Phoenix neurologist.

Five days later, the neurologist began his examination with questions:

"What year is this?

"What city are we in?

"Who is the President of the United States?

"Starting with the number one hundred, count backward by sevens."

David hesitated a couple of times during the recitation, but the doctor stopped him at seventy-two and gave him a pass. Next, the doctor watched him walk away and back and then tested his muscle strength, reflexes, coordination, vision, and hearing. He detected a lack of fine-touch sensation in both legs but found nothing that could exclude him from working.

The medical report was available two days later. David drove to the PFD and laid a copy on the chief's desk. Dave called him later that day and told him to report for duty on Christmas Eve, twenty-nine days after his horrific accident in the Sea of Cortez.

On Christmas Eve, an alarm sounded at 8:00 p.m. while David organized his equipment at the firehouse. Everyone gathered up their gear and hustled to their assigned vehicles. The big rigs rolled out to a large structure fire. The captains assessed the situation and planned their strategy. It was what firemen called a "working fire" that could easily spread to other structures. Off-duty firefighters, including Dave, were called in to assist. The pumper pulled up close, and the crew sprayed water within minutes of arrival. Other firefighters hustled to drag and connect more hoses to hydrants, but the flames licked high into the air and, at first, spurned the firefighting efforts.

David was always charged up when in action. He did his best to be in the middle of everything. It was nearly midnight before the firefighters had prevailed. There wasn't much left of the building, but nearby structures were unharmed. The flames finally surrendered with a last gasp, a tremendous mass of black smoke curling up into the night sky. The excitement over, David was tired but pleased that he was able to deal successfully with the considerable physical challenges as well as ongoing coordination issues.

* * *

He pushed himself to get stronger. Increasing his endurance proved to be the most difficult. He could walk at a good pace for an hour, but he could jog only a couple of blocks. Last year, he had joined the firemen's basketball team and had become one of the top scorers on the team. The team had already started practicing but enthusiastically welcomed him back as he came onto the court. David was apprehensive about what might happen, and things didn't go well. He couldn't keep up with the others or hit a basket — not a single score. Once, as the play changed directions, his feet got tangled, and several desperate steps kept him from falling on his face. The others understood. They offered words of encouragement, words that made him feel as if he were being pitied. Others, resorting to insider humor, ironically suggested he was never any good in the first place. One of his friends sat down next to him on the bench and reminded him that a few weeks ago he almost died and a full recovery wasn't going to happen overnight.

He slapped David on the knee and said in earnest, "Don't ever stop trying."

* * *

A local ENT doctor had been checking progress with David's right ear, and on January 5, 1983, he declared it fully healed and removed the myringotomy tube. A shocking loud "ping" echoed as the delicate hearing membrane recoiled in a dizzying explosion, apparently a normal response to what David thought would be "a piece of cake."

After several sessions of poking, grinding, and drilling, his dental work was finally finished as well.

He worked hard on his stamina and saw his exercise regimen pay off. He continued to reach new plateaus on a weekly basis. Thankfully, as he regained strength, his coordination was noticeably better as well. Two months later, he was handling ambulance and fire calls with no fatigue.

"We were all amazed at David's recovery; he is an amazing guy," Dave remembered. "A few months before the accident in Mexico, he and I were on rescue duty. A deluge in Prescott caused a lot of flooding. We responded to a call that someone was stuck in a collapsed basement. When we arrived at the location, a flooded stream blocked our way.

"First, we had to get a safety line across to the other side. A boy ran toward us on the other side of the stream, which was turbulent and loaded with debris. I was halfway across with the rope. In spite of our shouts, he jumped in and was quickly swept away. I lunged and grabbed him, but the river got me, too. I finally grabbed a signpost and hung onto it with one arm and to the boy with the other.

"David and a guy named Gene pulled us out, me and the boy. The firemen finally reached the crumbling house and rescued the boy's grandmother from the basement, which was filling up with floodwater. They clung to the rope as they carried her across the river.

"That's the way rescues were. We had to deal with whatever we faced, and we always had each other's backs."

***

David's problem finding the right word barely bothered him anymore. He had adjusted himself to the pace of his progress, and knowing he continued to get better with time, he accepted his situation with a deep sense of gratitude. He wasn't taking anything for granted. He had become grateful for his life, a feeling reflected in his face; his smile was always just waiting to happen.

Winter passed, and a lovely spring set in, the time to resume his outdoor activities. He started mountain biking again with his friends. He resumed climbing on Granite Mountain. He often thought about scuba diving, but when he did it was like reopening a nasty wound.

For the first time in his life, fear had taken hold over him. On waking early in the morning, he often found himself reliving over and over the absolute helplessness he had felt when he couldn't deflate the BC and the realization that, in a matter of seconds, his life had nearly been swept away from him. He remembered the excruciating frustration as he struggled to regain control over his mind and body.

Dr. Neuman's warning would echo inside his head: ". . . never dive again."

Then there were the less restrictive words from Dr. Adkisson. "When you're back to normal, I know of no scientific evidence that you wouldn't be able to dive again."

Was he completely back to normal? Which doctor was right?

David had experienced fear in the past, and he had developed his own method for dealing with it. Fear was present whenever he climbed a vertical cliff or kayaked through big rapids. Fear had to be respected but not allowed to subdue him. It had to be confronted using a combination of preparation and grit.

A vertical climb is a matter of one hold, one move, one step at a time. Continue balancing and repeating. Be patient, and do not give up. He visualized himself doing this before making his way up the nose of Yosemite's El Capitán. And it was not unusual for him when faced with the fear of a challenging ski slope or churning rapids, to persevere and push through it. The process of dealing with fear for David was becoming routine.

The only question left was quite specific. "How do I deal with the scuba diving problem?"

The answer came in the form of an article in an adventure magazine proclaiming the wonders of Hanauma Bay, with pictures and descriptions that stressed everything to love about scuba diving. One of the best diving sites in the world was a few miles south of Honolulu.

He needed a dive partner, and no question that would be George Calloway, an experienced diver. George was an adventurer, always interested in exploring new places. David tried to sound casual when he called him.

Aware of his friend's accident a few months before, George asked, "Are you up to it, David?"

"I'll be fine."

Life was going well for George. A bachelor, he had met a young woman with a recent and enthusiastic interest in scuba diving. He had begun practicing with her in the YMCA pool. Now she was eager to have her first dive in open water. David's plan was off to a perfect start.

He drove to a dive shop in Phoenix and invested in a state-of-the-art BCD with no less than four ways to release air from the bladder. Before leaving, he noticed a high-tech wrist-mount depth gauge and purchased it. Back home, he spread out his equipment and checked everything carefully. His first dive and ascent would be in a beautiful place, Hanauma, and the equipment wouldn't be worn until then.

He rode with George and his student to Sky Harbor Airport in Phoenix. George reminisced about the best adventures he and David had had together.

David smiled and listened, nodding his head in agreement, but like bees buzzing around his head, the tormented memories of the Sea of Cortez wouldn't leave him.

Hanauma Bay, Hawaii

After a six-hour flight, they landed in Honolulu, picked up a rental SUV, loaded up their gear, and headed south past Pearl Harbor and out of the busy city to a motel north of the bay.

They got up early the next morning and set out for the beach. Even from the shore, they saw the colors and shapes of the reef, the movement of the water turning them into a glistening abstract image. Dozens of snorkelers waded into the warm, clear water, lazily floating on the surface, breathing tubes protruding upward so the bodies looked like tiny submarines with periscopes raised above water, as they peered downward, hypnotized by the beauty below.

George first helped the woman get into her gear, reviewing safety instructions and underwater hand signals. He periodically cast a watchful eye in David's direction.

They waded in from the beach, gliding alongside snorkelers before the divers submerged into deeper water. They located "The Slot," a man-made (by dynamite in the 1950s) opening in the reef to accommodate a telephone cable lying flat on the ocean floor as it entered the bay from the open sea.

It was nearly impossible to swim through against the current, so the divers' choices were to swim over the reef or pull themselves through The Slot using the cable, hand over hand, to reach the far end. On this morning, the flow of the current was out to sea. They positioned themselves, one by one, in the current to be whisked out into the deeper water, a maneuver the locals called "shooting the slot." David's stomach churned.

George went first, holding his position in the middle of the current, glided through, and waited at the far end. It was a full minute before the young woman gathered the courage to stretch out and go with the flow, which she actually accomplished gracefully. Usually the lead diver, David had stationed himself at the rear, cautiously feeling his way along the course. The current swept him through the slot with ease, but the feat, which ordinarily would have given him a rush, was almost mechanical. His mind focused on the task ahead, surfacing from the depth.

He swam along the ocean floor for some time before signaling to George that he was ready to surface. At David's insistence, the two divers had an understanding. David would make the ascent on his own, while George monitored from below. George watched closely as David released air from his BCD, checked his depth gauge, and began a slow ascent. Getting to this point had been mechanical, but now he had to confront his fear — the mental thing. The ascent would be a face-to-face encounter with a demon.

George watched as David continued a slow, purposeful upward trajectory and made his way toward the shimmering sunlight overhead, monitoring the depth gauge and carefully measuring his safety stops.

Before he knew it, he found himself bobbing in the waves at the surface. It felt like a dream — surreal. And then it all went wrong.

The dream became a nightmare. He felt the dreadful foaming sensation in his legs, exactly as before. His chest tightened, and he struggled to breathe.

"Oh my God, I've done it again," he thought, "and this time I've killed myself!" He closed his eyes and surrendered himself completely to a will far greater than his own, just as in the Sea of Cortez.

But he hadn't blacked out. He opened his eyes to see the sun, bright and warm. He floated peacefully on the surface of the water, breathing easily. The foaming sensation in his legs and the tightness in his chest had cleared as quickly as they had overtaken him. The overwhelming sense he was going to die caused a psychosomatic event so real that David thought it would kill him.

Now a sense of euphoria overcame him, and he laughed. Floating alone in the middle of Hanauma Bay, he couldn't stop laughing.

He deflated the BCD and dived back down to where he could see George giving him the "OK" sign as he stood on the ocean floor. Then as he moved along with the others over gardens of coral, David's senses opened wide to the beauty around him. Schools of fish hovered over their favorite niches in the reef — parrotfish, raccoon butterflyfish, and damselfish. At one point, as he swam along the ocean floor, he was completely enveloped by an enormous school of large "humuhumu," triggerfish, the state fish of Hawaii, which had white stripes sweeping from the corners of their eyes to their tail fins. The undersea world he had known so well before had welcomed him back.

The ascent back to the beach was gradual. When they reached shallow water, they removed their fins and trudged onto the sand, where they sat in a circle. The girl put both arms around George's shoulders, kissed him on the cheek, and said, "That was the coolest thing I've ever done!"

David nodded his head and grinned, "Me, too." He had affirmed the idea that irrational fears can be vanquished by confronting them.

He winked at the both of them and said again, "Me, too."

Epilogue

David remained a member of the Prescott Fire Department until his retirement as a fire captain in 2004. Because of his mountaineering and whitewater skills, he organized and led a technical, high-angle and swiftwater rescue team for more than twenty years. He spent fifteen years as an instructor in wilderness emergency medicine at Prescott College, where he is now a member of the emeritus faculty.

Routine dental visits would affect his life importantly when Beth Klietz, his hygienist, accepted his request for a date. She already knew he was a fireman, and at every session she looked forward to reports of his latest adventures. From her reaction, he had found someone with a kindred spirit. Having grown up in Chicago, Beth hadn't been exposed to activities such as mountain climbing, but that changed quickly. When he told her about his diving accident, she was convinced God had a special purpose in mind for this man. In 1998, they were married.

Beth shares not only the household obligations but also works at his side in their real estate ventures and has become his partner in outdoor activities. A few years ago, they completed a 400-mile, 22-day, kayaking and camping trip from Prince Rupert, British Columbia, to Juneau, Alaska, coping with bears, mosquitoes, and rocky shores.

After stowing all their gear in Juneau, he teased, "Are you still going to stay married to me?"

"Doubt if I could ever find anyone better," she answered.

David's prodigious outdoor activities would be out of reach for most of us. They include climbing the Nose of El Capitán in Yosemite; kayaking the remote Cotahuasi River in Peru as a member of the second documented group to do so; skiing some of the steepest mountains in North America, often after climbing them in his ski equipment first; and scuba diving all over the world, often in the company of George Calloway.

When a massive earthquake struck the already challenged country of Haiti in January 2010, 230,000 people lost their lives, and many more were injured.

Without hesitation, David joined Dr. Jim Lindgren, a local emergency medicine doctor and founder of Window of Hope, a nonprofit medical and humanitarian organization providing assistance to vulnerable communities and disaster victims around the world. David's fellow paramedic at the Prescott Fire Department, Paul Williams, made it a group of three. "The trip to Haiti with Jim and Dave was life-changing for me," Paul remembered.

They booked a flight to Santo Domingo, Dominican Republic, then made their way to the Haitian border, where they found roads closed and severe gasoline shortages. Another organization, Mission of Hope, helped them get through the devastation to arrive at Port-au-Prince, where the trio of medics set up a clinic, examining patients on tables with sheets hung over lines for privacy.

"It was like pictures I'd seen of the Civil War — bodies strewn everywhere," David said. After treating numerous victims, they hitched a ride back to Miami on an Air Force C-130 and returned to Prescott.

"It felt like it was just something I needed to do," David said.

David's inquisitive nature culminated in a life-changing revelation. A voice, this time more of an intuition, confirmed to him that we live by faith not by sight. Again, he had a choice.

This time, it was a choice that we all have. He chose Life, according to John 14:6 NIV.

David Scalia kayaking the Cotahuasi River in Peru in 1995, only the second time it had ever been done. *(Photo courtesy of David Scalia)*

Selected Bibliography

Cousteau JY. First tests of the Aqua-Lung [Internet]. 1942 [cited 2015 Oct 2]. Available from: https://youtu.be/jgXGeZzzHD8

El-Orbany M, Ramez Salem M. Endotracheal tube cuff leaks: causes, consequences, and management. Anesth Analg 2013; 117:428-34.

Greyson B. The near-death experience scale: construction, reliablility, and validity. The Journal of Nervous and Mental Disease 1983; 171:369-75.

Harris W. How Gold Works. InfoSpace LLC [Internet]; 2009 Feb 23 [cited 2015 Oct 03]. Available from: http://science.howstuffworks.com/gold.htm

Appendix A

Buoyancy Control Devices

Fish have swim bladders that allow them to control their buoyancy by forcing air into or out of their anatomical "balloon." A buoyancy compensator device (BCD) is a piece of diving equipment worn to control underwater buoyancy by adjusting the volume of air in a bladder. The effect is the same as the fish's swim bladder. The diver adds air to the BCD through an inflation valve, and a vent valve allows the gas to be released. The diver controls the flow of air, in or out. With proper buoyancy, the work of swimming underwater is greatly reduced, just as it is for the fish.

Buoyancy is the trickiest part of diving. It changes significantly every time the diver changes depth. Both the Divers Alert Network (DAN) and Australia and New Zealand (ANZ) surveys found buoyancy problems were the most common adverse events leading to death. Victims often drowned because they were overweighted.

Wetsuits are important factors in buoyancy as well; their own buoyancy must be offset by a weight belt.

Forty percent of divers who perished were found to be grossly overweighted at the surface. As a diver descends, the wetsuit is compressed, and its buoyancy decreases. To restore balance, the BCD is inflated. Overweighting at the surface is considered by experts to be a dangerous practice that causes the diver to rely too much on the BCD.

Air must be released from the BCD during the ascent because the gas in the wetsuit as well as in the BCD itself will continue to expand, propelling the diver upward.

Recently, US doctors have suggested that drowning as the cause of death in 70% of scuba diving deaths may be an oversimplification, inappropriately attributing deaths to drowning because they occurred in the water. Sadler et al. cite two cases of sudden cardiac death as the cause of death and emphasize the inclusion of witness statements, equipment analysis, and autopsy findings to avoid mischaracterizing sudden natural death in the water as drowning.

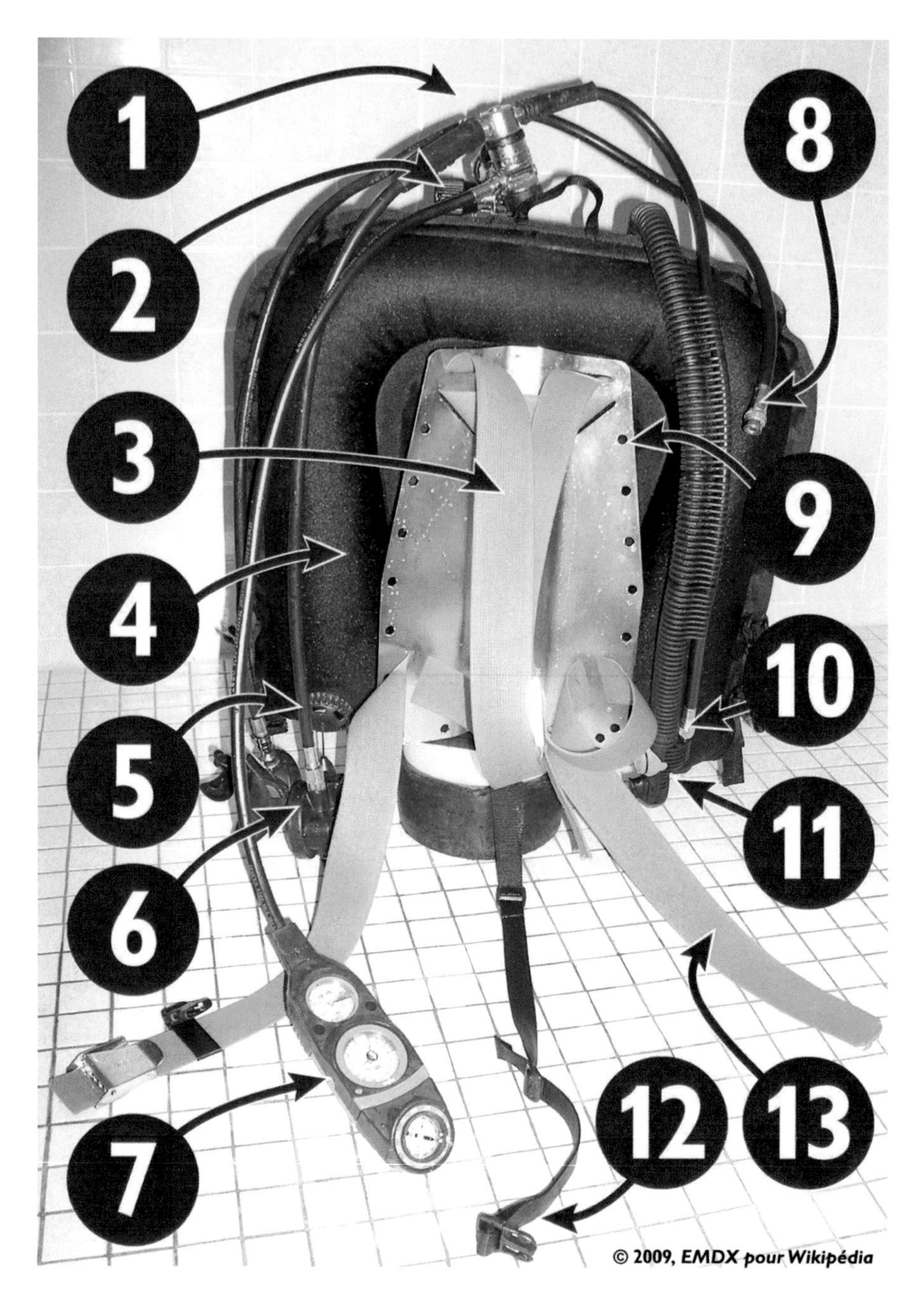

Buoyancy compensator device: 1) regulator first stage, 2) cylinder valve, 3) shoulder straps, 4) buoyancy compensator bladder, 5) relief and bottom manual dump valve, 6) regulator second stage (with "octopus"), 7) console (pressure gage, depth gauge, and compass), 8) drysuit inflator hose, 9) backplate, 10) BC inflator hose, 11) oral inflation mouthpiece and manual dump valve, 12) crotch strap, 13) waist strap. (Note: This is not a "horse collar" design.) *(Photo courtesy of Wikipedia)*

Hazards of BCDs include catastrophic failure due to punctures or tears that result in inadequate buoyancy to make a safe ascent. Malfunction of the inflator valve can result in uncontrolled ascent with associated risk of decompression sickness. Malfunction of the vent (or dump) valve, as occurred in David's case, may also result in an uncontrolled ascent. Newer BCDs have an overpressure release valve that automatically vents the bladder to prevent the rare but serious problems of rupture of the bladder and loss of buoyancy.

SELECTED BIBILIOGRAPHY

"Buoyancy compensator (diving)." Wikipedia: The Free Encyclopedia. Wikimedia Foundation, Inc., 28 June 2015 [cited 2014 Aug 30]. Available from: https://en.wikipedia.org/wiki/Buoyancy_compensator_(diving).

Holm J. Buoyancy control devices. Email to Monte Anderson, MD, 2014 July 28.

Pollock NW, Dunford RG, Denoble PJ, Dovenbarger JA, Caruso JL. Annual Diving Report, 2008 ed. Divers Alert Network: Durham, NC; 2008.

Professional Association of Diving Instructors (PADI). Buoyancy Control Device (BCD) [Internet]. Available from: https://www.padi.com/scuba-diving/padi-courses/about-scuba-gear/buoyancy-control-devices-(BCD)/

Sadler CA, Nelson C, Grover I, Witucki P, Neuman T. Dilemma of natural death while scuba diving. Acad Forensic Pathol. 2013; 3:202-12.

Appendix B

Scuba Diving Physiology

Atmospheric gas is 78% nitrogen and 21% oxygen, with small amounts of water vapor, carbon dioxide, argon, and other gases in trace amounts. Oxygen is mostly bound to hemoglobin in the bloodstream and is indispensable for many metabolic processes in the body.

Nitrogen is an inert gas; it is not metabolized in the body but dissolves in the bloodstream and is distributed to body tissues, moving from areas of high pressure to those with lower pressure. The presence of nitrogen can lead to serious problems with the pressure changes that occur in diving, especially when nitrogen bubbles into the bloodstream.

More than 100 years ago, Sorbonne Professor Paul Bert observed, "All symptoms, from the slightest to those that bring on sudden death, are the consequences of the liberation of bubbles of nitrogen in the blood, and even the tissues, when compression has lasted long enough.

"The great protection is slowness of decompression," he continued. These thoughts from "the father of pressure physiology" in 1879 remain valid today.

So why is there so much nitrogen in air? Air is the major reservoir for nitrogen, which eventually will form life-sustaining nitrogen compounds such as proteins. But humans cannot use nitrogen from the air. It first has to be broken apart (there are two atoms of nitrogen linked together in nitrogen gas), and the human body cannot do that. Only certain organisms, lightning, and industrial processing can.

Lightning breaks nitrogen molecules, enabling their atoms to combine in the air, forming nitrogen oxides that are carried to the soil by rain and snow. Certain bacteria and archaea can fix nitrogen, which then enters into plants that use inorganic nitrogen from soil to produce protein for their own nutrition, as well as ours, when we eat them. Finally, under great pressure and high temperature, industrial fixation combines atmospheric nitrogen with hydrogen to form ammonia and its compounds.

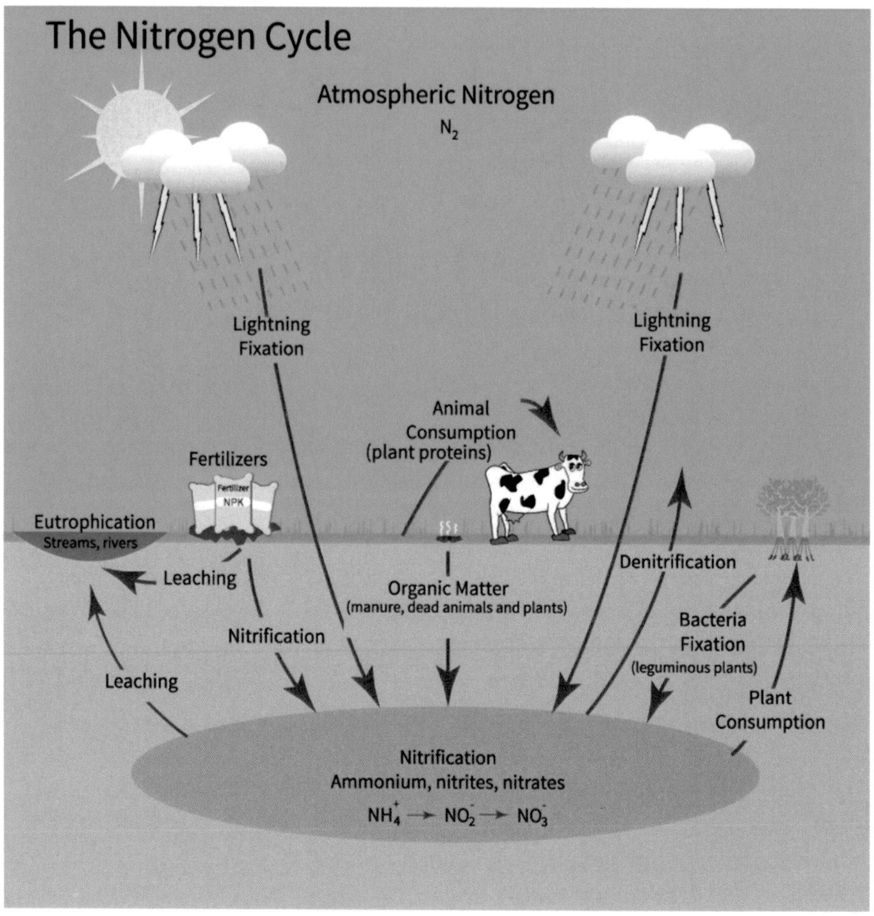

The nitrogen cycle *(Photo by Doethion/Adobe Stock)*

When the nitrogen gas pressure reaches a certain level during a dive — e.g., three atmospheres (at roughly 100 feet) — it may cause a feeling of euphoria (nitrogen narcosis) and impair the diver's judgment. It is relieved by ascending to a shallower depth.

In scuba diving, the pressure inside the body has to equal the ambient pressure (pressure outside the body). This equilibration is accomplished by the breathing gas regulator delivering pressurized gas to the lungs, which in turn deliver it to the bloodstream and finally into the tissues. As the pressure increases, increasing quantities of nitrogen dissolve into tissues, a process called on-gassing.

During the ascent, the process is reversed — off-gassing. An excessively rapid ascent can result in the formation of nitrogen gas bubbles that move from the higher pressure area of the tissues back into the bloodstream, where they are propelled to the lungs. An overwhelming mass of bubbles may damage the delicate alveoli (tiny lung sacs), thus creating a shunt that allows the nitrogen gas bubbles to enter the arterial circulation and go directly to the brain. The bubbles become arterial gas emboli, always life-threatening but much less frequent than decompression sickness, where there is no formation of emboli.

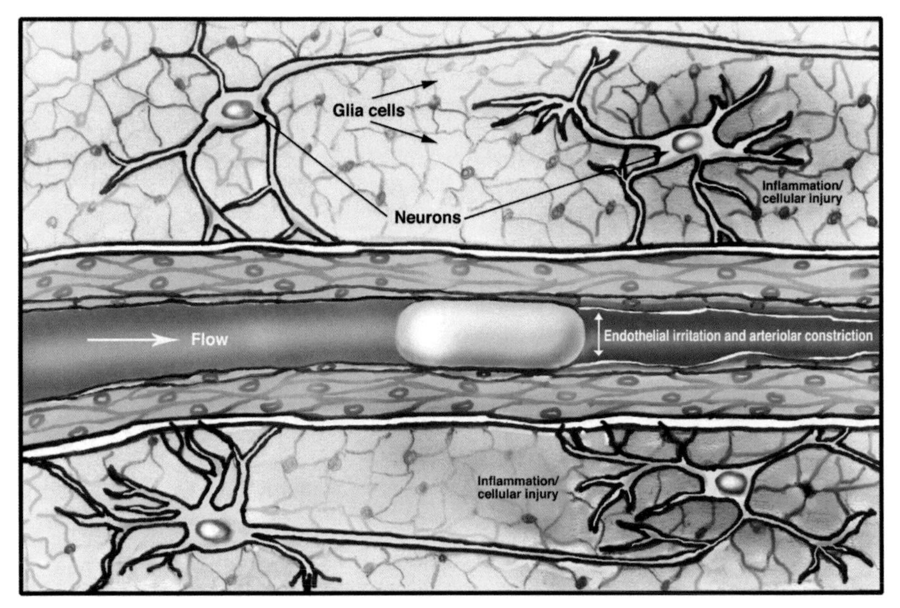

A gas bubble lodged in a small brain artery. Spherical bubbles typically assume a sausage shape as blood flow propels them into increasingly smaller vessels until they become lodged. This causes swelling of neighboring cells and constriction of the artery. Several other arteries may also be involved. An inflammatory response is triggered, and blood flow distal to the bubble ceases. Treatment in a hyperbaric chamber may decrease the size of the bubbles, making it easier for tissues to absorb them. At the same time, the blood still circulating is hyperoxygenated. *(Diagram by the author)*

All of the air breathed during a dive passes from the air tank through a scuba regulator. The regulator does two things. It reduces the pressure from the oxygen tank from 3000 psi down to 104 psi. In a second stage, it further reduces the pressure to ambient water pressure, which increases with the depth of the dive. The solubility of a gas (e.g., nitrogen) is proportional to its partial pressure (that of nitrogen alone) on a liquid (e.g., blood); thus the deeper one dives, the more nitrogen is dissolved in the blood and body tissues (Henry's Law).

As we have seen, off-gassing is the point where things can go wrong. US Navy dive tables specify time limits at various depths to prevent serious injuries. Bubble formation during a rapid ascent has been likened to the fizz when a carbonated beverage container is opened. In the moments before David became unconscious, he was aware this was happening to him.

During a normal (slow) ascent, nitrogen gas reenters the circulation and is transported to and then breathed out through the lungs. Ultrasonography has shown the existence of "silent" or asymptomatic bubbles that do not cause any apparent harm, the lungs being capable of metabolizing small bubbles.

But throughout the body, bubbles may form and enlarge in tissues as well as in the bloodstream. Extravascular bubbles tend to form mostly in areas where gas concentrations are highest — joints, tendons, and muscle sheaths. These areas tend to become painful in decompression sickness (DCS). Bubbles may also form in the central nervous system because nitrogen is highly soluble in fatty tissue, such as the myelin sheaths that surround and protect nerve cells.

Pulmonary decompression sickness (chokes) is a rare but serious form of DCS that occurs when an overwhelming stream of bubbles in lung capillaries virtually blocks the flow of blood through the lungs. The result is back pressure to the right ventricle (pulmonary hypertension) and inflammation of the small lung vessels followed by leakiness and pulmonary edema.

Bubbles in the arterial system are relatively rare but an extremely dangerous situation because the filtration process through the capillaries of the lungs has been bypassed. For bubbles to enter the arterial circulation, a shunt from the venous to the arterial circulation has to exist. Many people have a patent (open) foramen ovale (a hole in the heart that didn't close the way it should have after birth), which creates a congenital shunt that allows flow from the right side of the heart directly into the left side of the heart and out the aorta.

In diving injuries, the most common cause of a shunt is breath-holding during the ascent. The lungs expand rapidly with the decreasing ambient pressure to a critical point at which alveoli rupture and allow the gas bubbles to cross directly to the pulmonary vein that directs flow back to the left side of the heart and out into the aorta. Whether bubbles enter the arterial system via a patent foramen ovale or ruptured alveoli, the result is arterial gas embolism (AGE). The effect is the same as any other form of embolism, whether it is due to a blood clot, plaque, or air embolism. Arterial emboli can lodge in coronary arteries as they flow out through the aorta (rare) or into the brain via the carotid arteries. Distal to the point of arterial obstruction, cell damage and cell death occur quickly.

The following steps are the same in the treatment of DCS or AGE:

- Check airway.
- Check breathing.
- Check cardiac (heart).
- Place in supine position (on the back): "supine is fine."
- Administer 100% oxygen, the sooner the better.
- Do not attempt in-water decompression (taking the patient back to a lower depth).

- Call the Divers Alert Network (DAN) 24-hour emergency hotline (+1-919-684-9111), which provides expert help with diving emergencies.
- Get to the closest emergency room.
- Recompress in a hyperbaric chamber.

Undersea and hyperbaric medicine specialists point out that oxygen is their drug, and pressure is what they use to "crunch bubbles," to shrink them to a smaller size and eventually get the nitrogen back into solution. A pressure of 3 atmospheres (3 ATA) will shrink the total bubble volume to one-third of its original size, a remarkable therapeutic benefit as illustrated in David Scalia's story.

SELECTED BIBILIOGRAPHY

Branger AB, Eckmann DM. Theoretical and experimental intravascular gas embolization absorption dynamics. J Appl Physiol. 1999 Oct; 87(4):1287-95.

"Decompression theory." Wikipedia: The Free Encyclopedia. Wikimedia Foundation, Inc., 2015 Aug 29 [cited 2015 Aug 18]. Available from: https://en.wikipedia.org/wiki/Decompression_theory

Freudenrich C. How Scuba Works [Internet]. HowStuffWorks, LLC; 2001 May 29 [cited 2015 Oct 03]. Available from: http://adventure.howstuffworks.com/outdoor-activities/water-sports/scuba3.htm/printable

Kellogg RH. La pression barométrique: Paul Bert's hypoxia theory and its critics. Respir Physiol. 1978 Jul; 34(1):1-28.

Lynch JH, Bove AA. Diving medicine: a review of current evidence. J Am Board Fam Med. 2009 Jul-Aug; 22(4):399-407.

Martin L. Scuba Diving Explained: Questions and Answers on Physiology and Medical Aspects of Scuba Diving. Flagstaff, AZ: Best Publishing Company; 1997.

Muth CM, Shank ES. Gas embolism. N Engl J Med 2000 Feb; 342(7):476-82.

Pidwirny M. The nitrogen cycle. Fundamentals of Physical Geography, 2nd ed. [Internet]. 2006 [cited 2015 Aug 18]. Available from: http://www.physicalgeography.net/fundamentals/9s.html

Smith RM, Neuman TS. Elevation of serum creatine kinase in divers with arterial gas embolization. N Engl J Med. 1994 Jan 6; 330(1):19-24.

Appendix C

Decompression Chambers and Hyperbaric Medicine

Decompression chambers were invented in 1916 by Italian engineer Alberto Gianni specifically to expedite divers in classic diving suits with big helmets and surface-supplied air through a long hose from a ship. When divers were in cold or dangerous situations, they could ascend quickly and enter the decompression chamber on a ship, for example, and equilibrate on the surface.

The term "decompression chamber" is misleading. Recompression using hyperbaric pressure to reproduce the hyperbaric conditions under the sea is what actually occurs. At the beginning of the treatment, the patient is said to be descending (exactly as if diving); treatment ends with a controlled ascent back to normal atmospheric pressure even though the subject is in a stationary chamber that doesn't move at all.

During the Industrial Revolution, people turned from wood to coal as a prime energy source. Much of the European coal lay beneath boggy terrain, mud, and quicksand. Along came caissons, large structures that resembled modern silos. They were positioned so they could be settled through the muck to a seam of coal somewhere beneath the surface. Workers climbed down ladders to the bottom of the gigantic tubes, where they would scoop out debris, shoveling it into large buckets hauled to the top of the tube with ropes and pulleys and dumped. Once a coal bed was located, it was extracted in the same fashion as the muck.

Caissons were pressurized to keep the water outside from seeping in. Pressures may be equivalent to immersion in considerable depths of seawater. Symptoms of severe back pain might cause a caisson worker to assume a forward bent posture that came to be known as "the bends." Associated symptoms included pain in multiple joints, epigastric pain, paralysis, and even death. First called "caisson disease," identical symptoms were later seen in scuba divers and aviators. It was the same entity — decompression sickness.

The concept of caisson technology captured the imagination of bridge builders everywhere who at first ignored the associated medical problems. James Eads, the designer of the Eads Bridge across the Mississippi River, became concerned about the health of the workers. He asked his friend, Dr. Alphonse Jaminet, to investigate.

In 14 months, the doctor treated 119 cases of severe decompression sickness. Fourteen workers died. On one of his descents into a caisson, Jaminet himself became paralyzed and aphasic on exiting the airlock. Fortunately, he later recovered.

Work began on the Brooklyn Bridge of New York's East River in 1870. John Roebling, who had been a principal designer of the project, died during the construction. His son, Washington Roebling, assumed supervision of the project. The young Roebling knew about the medical problems at the Eads Bridge and engaged Dr. Andrew Smith as an on-site physician. In five months, Smith faced 110 cases of decompression sickness and 2 deaths. As a result of a visit inside one of the caissons, Roebling himself was paralyzed and never fully recovered.

Both Jaminet and Roebling made recommendations: more rest between shifts, proper diet, regular sleep, and avoidance of alcohol and tobacco. Both doctors recommended installation of elevators in place of stairs in caissons, but there was a long delay before implementation.

Serious problems continued to occur, usually within an hour of leaving the caisson. After one to three days, most patients' symptoms would resolve on their own. Smith recommended a recompression chamber be placed onsite, with treatment to be started immediately on presentation of symptoms. That didn't happen, but the elevators were eventually installed.

Eads Bridge under construction, 1873 *(Photo courtesy of Library of Congress Prints and Photographs Division, Washington DC, 20540; photo by C.M. Woodward)*

Caissons were sunk in place by seven-ton blocks on the roofs of the structures. When sand was reached, men entered the air chambers through airlocks and the main shaft. At the bottom of the caisson, they excavated the riverbed by hand in hot, humid conditions, shoveling sand into a pump that suctioned it out to the top of the structure. Elevators (upper left panel), recommended by physicians, eventually replaced stairways. Compressed air was pumped into the chamber to keep water out, leading many to suffer caisson disease (the bends) due to the accumulation of nitrogen gas in the body. When bedrock was reached, the caisson was filled with concrete, and the foundation was fortified by filling the air chamber with concrete. As the men ascended to the surface, concrete filled the entire caisson behind them.

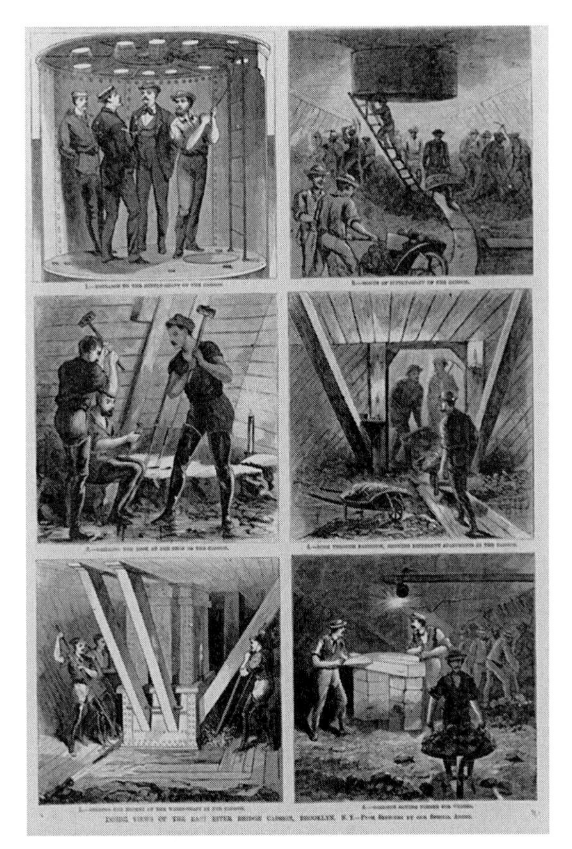

Inside views of the East River Bridge caisson *(Photo courtesy of Library of Congress Prints and Photographs Division, Washington DC, 20540)*

In the 1870s, Paul Bert described the physiological effects of varying air pressure — oxygen toxicity on the central nervous system — and suggested properly dosed oxygen should be used in the treatment of decompression sickness. "All symptoms," he said, "from the slightest to those that bring on sudden death, are the consequences of the liberation of bubbles of nitrogen in the blood, and even in the tissues, when compression has lasted long enough." He added, "The great protection is slowness of decompression."

In the 1930s, Dr. Albert Behnke of the US Naval Medical Corps and Louis Shaw were the first to use oxygen in this manner. They went on to identify safe time-dose exposure limits.

In the 1940s, Dr. Ita Boerema, a cardiac surgeon at the University of Amsterdam, needed to find a way to clamp off the heart long enough to successfully operate on difficult heart problems. He and his colleagues experimented with hypothermia, the goal being to "reduce the metabolism of a warm-blooded animal to such an extent that the physical processes would almost come to a standstill." The hypothermia doubled the ischemic time, but that meant only a period of five minutes, which was not nearly enough.

Using the information provided earlier by Behnke and Shaw, Boerema cooled and exposed dogs to 3.0 ATA (one atmosphere [ATA] is the air pressure at sea level), the level proposed as the upper safe threshold. It worked. Boerema drained almost all of the blood from pigs via exchange transfusions with a Ringer's-like solution. Hemoglobin levels were close to zero. For 45 minutes, the animals received oxygen via an endotracheal tube and breathed at a pressure of 3.0 ATA in a high-pressure tank.

"During all this time, the EEG showed no pathological changes, the circulation and blood pressure remained spontaneously normal," Boerema reported. "Recovery [of the animals] was uneventful after re-infusion of blood."

By 1959, Boerema and his group were successfully performing heart surgery on infants with congenital heart problems and adults inside a specially built hyperbaric operating room. For his bold and revolutionary advancements, Boerema has been recognized as "the father of modern hyperbaric medicine."

Many of Boerema's colleagues couldn't tolerate the environment inside a hyperbaric operating room, which would be replaced by modern heart-lung bypass machines in a standard operating room. The newer devices supported blood circulation and oxygenation for more than six hours, as well as induction of hypothermia and rewarming of the patient without requiring the high-pressure environment.

Entertainer Michael Jackson helped introduce hyperbaric chambers to the general public in the mid-1980s when a photograph of him in a hyperbaric chamber made the front pages of the popular media. Other health seekers continue to purchase, rent, or rent-to-own the devices with the delusive goal of holding back the aging process.

In clinical situations, hyperbaric equipment allows variable increases in air pressure, depending on the structure of the unit, usually two or three times normal but up to six times normal in some units. As already noted, using Boyle's law, at 3 ATA, the bubble volume will be reduced by approximately two-thirds. The patient receives 100% oxygen for established time periods via oxygen masks (including built-in breathing systems, or BIBS), transparent hoods that cover the head, or endotracheal tubes. Some larger units may treat more than one person at a time.

The National Fire Protection Association (NFPA) recognizes three categories of hyperbaric chambers. Class A (multiplace, more than one patient) units are required to have fire suppression systems and are constantly monitored to ensure that the oxygen percentage (PO_2) never goes above 25%. Class B (monoplace, just one patient) are much smaller and usually pressurized with an oxygen-rich environment (at or near 100%), making the risk of explosions and fire much greater than in Class A units. Class C units may be used only for animals.

In Class A units, as used in David's treatment, an inside attendant remains with the patient(s) in the chamber throughout the treatments, which may last from five to eight hours. The attendant's main task is to observe for seizures, difficulty breathing, etc. All treatment protocols utilize periodic rest periods away from the 100% oxygen for several minutes, and the inside attendant disconnects the hoods or other devices during these important pauses, which last five to ten minutes and reduce the risk of oxygen toxicity.

Staffing in Class A units includes the physician in charge, a chamber operator/outside observer, and an inside attendant/inside observer. The physician in charge assesses the patient and writes the orders. The chamber operator monitors and controls what gas (air or oxygen) is delivered to the patient from a control panel outside the chamber. He is in constant communication with the inside attendant by headphones.

For the duration of the treatment, the inside attendant is subjected to the same pressure as the patient(s) which, unlike the patients who are breathing oxygen and little or no inert gas, puts the provider at risk of DCS. When this happens, the attendant may need oxygen supplementation.

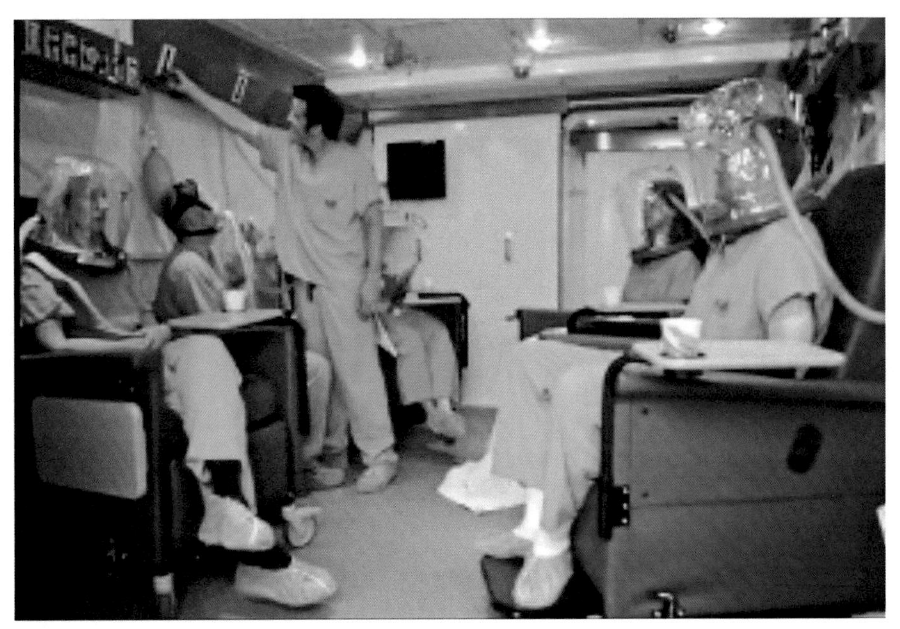

Multiple patients inside a multiplace Class A pressurized chamber wear oxygen hoods. The inside attendant (without oxygen hood) stands in the rear. *(Photo courtesy of Dr. James Holm.)*

Patients from the intensive care unit (ICU) need ICU personnel and respiratory therapists on hand during therapy in the chamber.

Class B, smaller monoplace chambers, are filled with 100% oxygen in the chamber itself. The risk of explosions and fire are much greater than Class A units because of the large volume of concentrated oxygen. These smaller units are available in some hyperbaric treatment centers because of their flexibility to deliver medical grade oxygen or medical grade air.

Indications (a condition that makes a particular treatment or procedure advisable) for hyperbaric oxygen therapy include arterial gas embolism caused most commonly by acute severe diving injuries and pulmonary barotraumas but also by mechanical ventilation, placement of central lines into large veins, and hemodialysis.

Other indications approved by the Undersea and Hyperbaric Medical Society (UHMS) are carbon monoxide poisoning, clostridial myositis, myonecrosis (gas gangrene), crush injuries, compartment syndromes, other acute traumatic peripheral ischemias, decompression sickness, enhancement of healing in select problem wounds, exceptional blood loss anemia, intracranial abscess, necrotizing soft tissue infections, refractory osteomyelitis, compromised skin flaps and grafts, delayed radiation injury, and thermal burns.

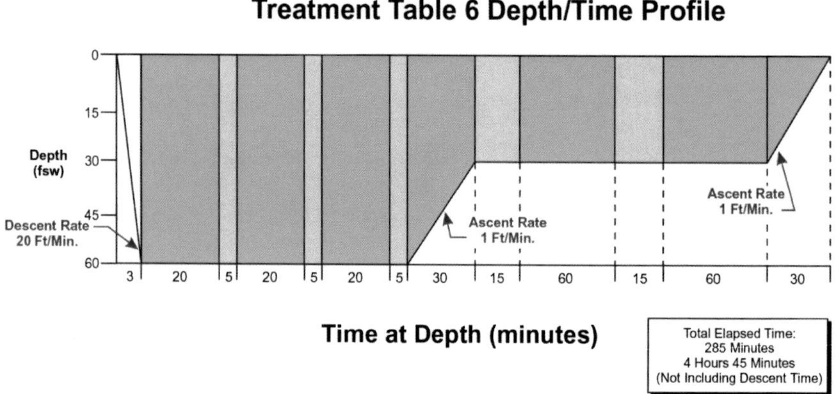

This schematic diagram is of the US Navy Treatment Table 6, one of the most commonly used tables for treating serious decompression sickness. Pressure increases with depth and decreases as it returns to depth zero. Note: fsw = feet of seawater. (*Photo courtesy of* US Navy Diving Manual)

SELECTED BIBLIOGRAPHY

Behnke AR, Shaw LA, Messer AC, Thomson RM, Motley EP. The circulatory and respiratory disturbances of acute compressed-air illness and the administration of oxygen as a therapeutic measure. Am J Physiol. 1936; 114(3):526-33.

Boerema I, Meyne NG, Brummelkamp WK, Bouma MH, Mensch F, Kamermans M, Stern H, Van Aalderen W. Life without blood; a study of the influence of high atmospheric pressure and hypothermia on dilution of the blood. J Cardiovasc Surg. 1960;1:133-46.

Butler WP. Caisson disease during the construction of the Eads and Brooklyn Bridges: A review. Undersea Hyperb Med. 2004; 31(4):445-59.

Gill AL, Bell CNA. Hyperbaric oxygen: Its uses, mechanism of action and outcomes. QJM. 2004 Jul; 97(7):385-95.

Holm J. Decompression chambers [online]. Email to Monte Anderson, MD. 2014 Jul 28.

Kellogg RH. La Pression Barométrique: Paul Bert's hypoxia theory and its critics. Respir. Physiol. 1978 Jul; 34(1):1-28.

Neubauer RA, Maxfield WS. The polemics of hyperbaric medicine. J Am Phys Surg. 2005; 10:15-17.

Pulley SA. Decompression sickness [Internet]. 2014 [updated 2014 May 9]. Available from: http://emedicine.medscape.com/article/769717-overview

Singh S, Gambert SR. Hyperbaric oxygen therapy: a brief history and review of its benefits and indications for the older adult patient. Ann Longterm Care 2014 Jul; 22(7):37-42. Available from: http://www.annalsoflongtermcare.com/print/2398

Appendix D
Saturation Diving

Saturation diving is a complex technology enabling teams of divers to spend prolonged tours of duty at great depths. Divers eat and sleep in a submerged capsule similar to a space capsule. The practice became possible when it was realized that, after hours under pressure, the body becomes saturated with inert gas and no further uptake will occur. A single, but prolonged, decompression at the end of the tour is all that is required.

In 1937, Max Gene Nohl, an MIT graduate, salvage diver, and adventurer, made a world-record 420-foot dive to the bottom of Lake Michigan while breathing a helium-oxygen gas mixture he developed along with Dr. Edgar End. The following year the two undersea pioneers made the first intentional saturation dive, spending 27 hours breathing air at the equivalent of 101 feet of water in the recompression facility in Milwaukee, Wisconsin. After several hours under pressure during these deep dives, the body becomes saturated with inert gas, and no further uptake occurs. Therefore divers may remain at depth for days, and a single decompression is all that is necessary at the end of a tour of duty.

In the 1960s, US Navy Captain George Bond became the senior medical officer and principal investigator for the navy's Man-in-the-Sea Program: Sealab I and Sealab II. In 1964, four navy divers were submerged 192 feet in the Sealab I underwater habitat in the Atlantic Ocean near Bermuda. The men worked six hours a day on the ocean floor outside of the habitat without returning to the surface. They were brought back to the surface after eleven days because of an impending tropical storm.

In 1965, Sealab II continued the pioneering work on the ocean floor. It was a 57-foot-long capsule that lay 205 feet deep in the Pacific Ocean off La Jolla, California. Teams of aquanauts, including former astronaut Scott Carpenter, lived and worked for days from an underwater module. They lived in a pressurized environment and would be decompressed only once at the end of their tour of duty (an average of 15 days, but Carpenter remained for 30 days). The men — including scientists, oceanographers, medical doctors, engineers, and photographers — worked six hours every day in murky water at 48-52°F, diving up to 100 feet lower than Sealab II itself. A large hose carried fresh water, as well as telephone and

Sealab I rests on a pier after submersion at 192 feet of seawater off Burmuda. *(Photo courtesy of Wikipedia)*

television connections, from the support ship above them. A trained porpoise named Tuffy also shuttled supplies to Sealab. The aquanauts breathed a mixture of 74.4% helium, 21.6% oxygen, and 4% nitrogen. Because of the helium their voices sounded a bit like cartoon chipmunks, but overall they fared well, and the teams completed a number of important scientific explorations that included harvesting food from the deep.

Since these early expeditions, more elaborate underwater habitats with several different chamber areas have been devised. The structures are not unlike space capsules; both are living chambers devised to allow people to exist in an otherwise impossible environment. In oceanography, a diving bell serves as an elevator between the habitat on the ocean floor and the surface as well as between the habitat and the work site. The underwater habitat functions like a space station and the diving bell like a space capsule. A support ship on the ocean surface has a "dive control" team that takes control once divers enter a bell to be lowered to the chamber below. The team monitors and controls gas pressure, mixture, and temperature. Bundled lines and cables, connecting the ship to the undersea structures, handle communications, inhaled and exhaled air, power, and warm water. The cable may be several hundred meters in length. When submerged, descending, and working, divers are connected to the diving bell by individual umbilical cords.

As the depths of ocean exploration deepened to greater than 500 feet (150 m), a serious problem called "high-pressure nervous syndrome" (HPNS) arose. Because of the condition nitrogen narcosis, which typically occurs at depths greater than 100 feet, new gas mixtures were being tested, including heliox, which is composed of only helium and oxygen. First described by Peter B. Bennett, HPNS is characterized by tremors of the hands, arms, and torso (at first called helium tremor), dizziness, nausea, vomiting, electroencephalographic abnormalities, and lapses of consciousness. It was clearly a different disorder than nitrogen narcosis, which does not involve tremors of any sort.

Believing the symptoms of HPNS could be controlled by adjusting the inert gas mixtures, Bennett experimented with different mixtures of nitrogen, helium, and oxygen (trimix). Inert gases had been ranked according to their narcotic potency in this order (from least to most): helium, neon, hydrogen, nitrogen, argon, krypton, and xenon. Interestingly, the potency is directly related to solubility in lipid (helium least and xenon most).

Between 1978 and 1984, the Atlantis dive series — experiments with humans breathing different inert gas formulations — were conducted in a seven-foot-wide steel wet chamber that could achieve simulated depths of 3600 feet (1097 m). In the Atlantis III experiment in 1981, three divers, under meticulous observation, successfully reached a depth of 2,250 feet, breaking the existing world record. The depth was achieved with slow rates of compression and decompression, seven days to reach the greatest depth and thirty days to resurface, illustrating one of the practical challenges in deep diving. On completion of Atlantis III, Bennett noted, "The experiments show that the [three young] men were fit and well and able to perform complex work for four days not only at 650 m (2,132 feet) but also during 24 hours at the greater depth of 686 m (2,250 feet) after the completion of the four days at 650 m. At 650 m, one of the divers performed work on the bicycle ergometer at 250 W for five minutes, and the arterial carbon dioxide was no higher than 53 mmHg (7 kPa). With the 10% nitrogen mixture, no dyspnea was reported by the divers."

Research continues to determine the exact causes of HPNS. Factors continuing to be studied are the following: variables that make some individuals more resistant to HPNS than others; rates of compression; the addition of anesthetics to the gas mixtures; and

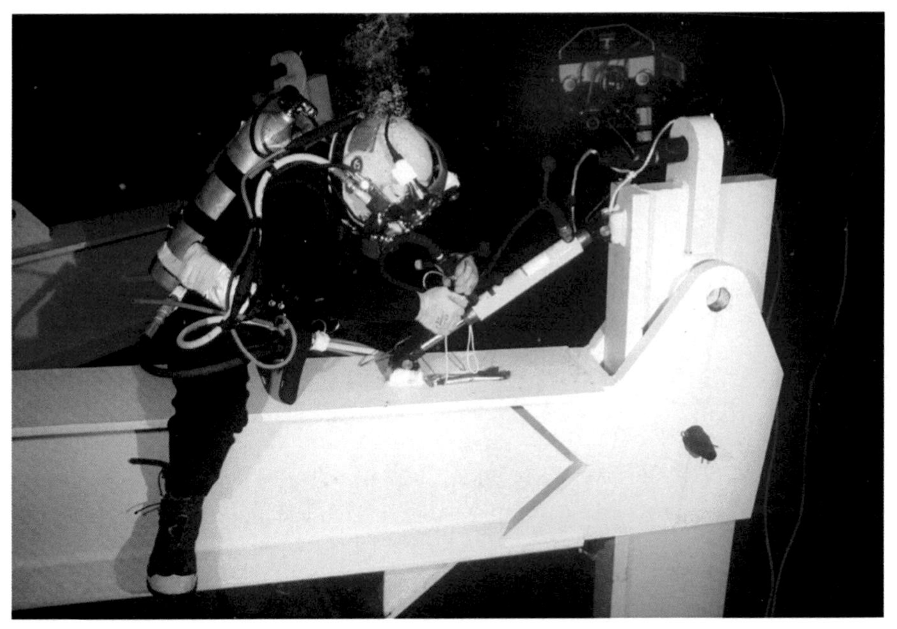

Saturation diver wearing a commercial diving helmet with breathing system, oral-nasal mask, defogger, headlight system, backup breathing gas system, communication system, head cushions, and water ejection system. The helmet weighs more than 30 lbs. Umbilical connections to the support station float behind the diver. *(Photo courtesy of US Navy)*

adaption after compression before starting work. There is also great interest regarding specific neurotransmitters like GABA (gamma-amino butyric acid) and depletion of substances like monoamines in the CNS that lower the threshold for seizures.

Using the knowledge of saturation diving accumulated over the past sixty years, the field of commercial diving came into being. Deep-sea saturation divers do undersea work in salvage operations and in the oil and gas industry.

SELECTED BIBLIOGRAPHY

Bennett PB, Coggin R, Roby J. Control of HPNS in humans during rapid decompression with trimix to 650 m (2132 ft.). Undersea Biomed Res. 1981; 8(2): 85-100.

Capt. George Bond, 67, dies; pioneer in underwater living. New York Times [Internet]. 1983 Jan 6. Available from: http://www.nytimes.com/1983/01/06/obituaries/capt-george-bond-67-dies-pioneer-in-underwater-living.html

Goldstein R. Scott Carpenter, one of the original seven astronauts, is dead at 88. The New York Times [Internet]. 2013 Oct 10 [cited 2015 Jan]. Available from: http://www.nytimes.com/2013/10/11/us/scott-carpenter-mercury-astronaut-who-orbited-earth-dies-at-88.html

Hong SK, Bennett PB, Shiraki K, Lin Y, Claybaugh JR. Mixed-gas saturation diving. Compr Physiol. 2011; 1023-45.

How Does Saturation Diving Work? [Internet]. HowStuffWorks, LLC; 2015 Jan [cited 2015 Oct 03]. Available from: http://science.howstuffworks.com/science-vs-myth/everyday-myths/question640.htm

Rostain JC, Balon N. Recent neurochemical basis of inert gas narcosis and pressure effects. Undersea Hyperb Med. 2006 May-Jun; 33(3):197-204.

Talpalar AE, Grossman Y. CNS manifestations of HPNS: revisited. Undersea Hyperb Med. 2006; 33(3):205-10.

Tilford EJ, US Navy, ID 020723-N-7479T-002 [Internet]. 2002 Jul 21 [cited 2015 Oct 03]. Available from: http://www.navy.mil/view_image.asp?id=2118

Vanderwerff S. An untapped ocean of opportunity, Part II [Internet]. 2014 Jul 31 [cited 2015 Oct 03]. Available from: navymedicine.navylive.dodlive.mil/archives/6587

Vanderwerff S. An untapped ocean of opportunity, Part III [Internet]. 2014 Aug 07 [cited 2015 Oct 03]. Available from: navymedicine.navylive.dodlive.mil/archives/6682

Appendix E

San Diego Reflections

USS *Dixon*

The USS *Dixon* has a story of her own. Built in 1970, it had 995 compartments that housed 1,100 sailors on 12 decks. The crew worked to meet up with the submarines, wherever they may have been, and offered medical care, made repairs, provided cleaning services, scraped barnacles, replaced parts, etc. The ship was named for George E. Dixon, commander of the Confederate submarine *H.L. Hunley*. Designed to provide logistical and technical support for up to 12 nuclear attack submarines, the *Dixon* also had the ability to moor up to four submarines alongside at one time.

USS *Dixon* at sea, photo signed by Captain Heuberger *(US Navy photograph courtesy of US Navy; photo from David Scalia's scrapbook)*

The ship was home to Dr. Greg Adkisson for a good part of his service in the navy, including a seven-month tour in the Indian Ocean. It always carried the decompression chamber that saved David Scalia's life.

In 1995, the *Dixon* sailed for the last time to Norfolk, Virginia, passing through the Panama Canal. On July 21, 2003, it was sunk in the Atlantic Ocean, along with two other navy ships, by a barrage of navy guns, missiles, and bombs, southeast of Charleston, South Carolina.

Dr. Adkisson, Dr. Neuman, and Dr. Phillips

Because of their specialized training, Dr. Greg Adkisson and Dr. Tom Neuman worked together to train resident physicians in undersea and hyperbaric medicine. Dr. Neuman ensured someone from his small informal group, which included Dr. Paul Phillips, was available for interaction with the navy personnel whenever a case came in. The collaboration is remembered as a cordial and beneficial arrangement lasting for several years until the UCSD Medical Center received its own decompression chamber and established the Division of Hyperbaric Medicine in 1984.

Lieutenant Commander Gregory H. Adkisson was 30 years old when he treated David Scalia. Born at an army post in Germany, he grew up as an army brat, living in a variety of locations, and graduated from high school in Lincoln, Nebraska. From there, he entered the US Naval Academy. During his freshman year, he decided to resign from the academy to study premed, something not offered at Annapolis. Shortly after submitting his resignation but prior to his actual separation, the academy announced a premed program offered on a trial basis. Dr. Adkisson applied for the highly competitive and limited program with little hope that he would be accepted. But a few weeks later, the admiral in charge notified him that he was one of 14 freshmen accepted. He graduated from the academy with a degree in bioscience and was one of six admitted to medical school that year.

(Photo courtesy of Greg Adkisson)

After medical school at the University of Arizona, he began an internship at the Naval Medical Center San Diego. Unsure of what field of medicine he would follow, and given his interest in the operational aspects of navy medicine, he applied for SEAL training. Although accepted to the program, he unexpectedly heard from his executive officer that they needed a "volunteer" to go to Guantamo Bay, Cuba, as an acting flight surgeon. Without much choice, he found himself a resident of the Naval Air Station in Guantanamo Bay. He completed training in undersea and hyperbaric medicine, eventually attaining qualification as a medical officer in submarines, diving medicine and saturation diving. *(See Appendix D: Saturation Diving.)*

During his time as an undersea medical officer, he also served as exchange officer in undersea medicine to the United Kingdom, where he became part of the Submarine Parachute Assistance Group, working with the Royal Navy for two years.

Looking back at his military career, he remarked, "At the time, it was hard to accept that someone behind the scenes, I don't know who, decided that I couldn't finish

SEAL training because it wouldn't look good for a doctor to do this. So I cooled my heels. Strange how things work out, because I ended up working with those guys (SEALS) later on. Eventually, I ended up doing all of the fun things you can do in the military."

Later in his career, Dr. Adkisson completed an anesthesia residency at the Naval Medical Center San Diego and attended the US Naval War College in Newport, Rhode Island, where he was named an honor graduate. As a captain, he held two commands, first as commander of the Defense Medical Readiness Training Institute at Fort Sam Houston, Texas, and then as the commanding officer of Naval Health Clinic New England, a seven-state regional command headquartered in Newport, Rhode Island.

Upon leaving the navy, he entered the private practice of anesthesiology. He continues a distinguished career and currently works as a regional director and startup chairman/mentor for a large anesthesia group based in New York.

A super-achiever, **Tom Neuman** attended Cornell University on a Regents' Scholarship and graduated from New York University School of Medicine, where he received the Lang Award as outstanding student in his class. He earned board certifications in internal medicine, pulmonary disease, occupational medicine, emergency medicine, and undersea and hyperbaric medicine. While in the US Navy in the 1970s, he completed training at the Naval School of Submarine Medicine, the Naval Diving and Salvage Training Center, the Naval Nuclear Power Training Unit, and the Naval School of Deep Diving Systems. He served as Submarine Medical Officer, Undersea Medical Officer, and Saturation Diving Medical Officer.

Dr. Tom S. Neuman, 1983 *(Photo courtesy of Tom S. Neuman)*

Dr. Neuman continued his academic career, becoming research director of the Hyperbaric Medicine Center at UCSD Medical Center, president of the Undersea and Hyperbaric Medical Society, and editor of the "bible of undersea medicine," *Bennett and Elliott's Physiology and Medicine of Diving*. The main focus of his extraordinary career has been undersea medicine, and he has authored many original scientific studies in the field.

Dr. Paul Phillips was a 28-year-old medical "whiz kid" who was into his final year of residency training when he crossed an unforgettable path with David Scalia. He had graduated from the University of Rochester, New York, with high honors and entered medical school there, again graduating with honors. He moved to San Diego in 1980 for residency training in internal medicine and was in his third and final year of training when he took that rescue flight to Mexico. Had it not been for this brilliant young doctor's arrival on scene, David would not have survived that crucial night.

Dr. Paul Phillips, 1982 *(Photo courtesy of Paul Phillips)*

As he had planned all along, Dr. Phillips completed specialty training in cardiology at UCSD Medical Center, Scripps Mercy Hospital, and Hospital Universitario San Carlos in Madrid, Spain. He became an interventional cardiologist and a dedicated teacher in the Division of Cardiology at UCSD and director of Scripps Health's heart catheterization laboratories. He is coauthor of *The Stenter's Notebook* and a contributor to many original articles in cardiology literature, including the first paper to show toxicity in patients with muscle complaints while on statin drugs and with normal muscle enzymes.

SELECTED BIBLIOGRAPHY

United States Navy Submarine Tenders, USS Dixon AS 37 [Internet]. 1997, 2005 [cited 2015 Oct 03]. Available from: tendertale.com/tenders/137/137.html

Appendix F

Barotrauma to the Head (Squeezes)

Middle-ear barotrauma, or "middle-ear squeeze," is the most common barometric injury in scuba diving. It results in acute pain and occurs in up to 30% of first-time divers and in as many as 10% of experienced divers. The Eustachian tubes connect the inner ear chamber with the back of the nose. With increased pressure on the middle ear, the tympanic membrane (eardrum) may be deflected inward to the point of rupture unless the pressure is balanced by external air pressure passing through the Eustachian tube. Obviously, head congestion and compromise to the Eustachian tubes will

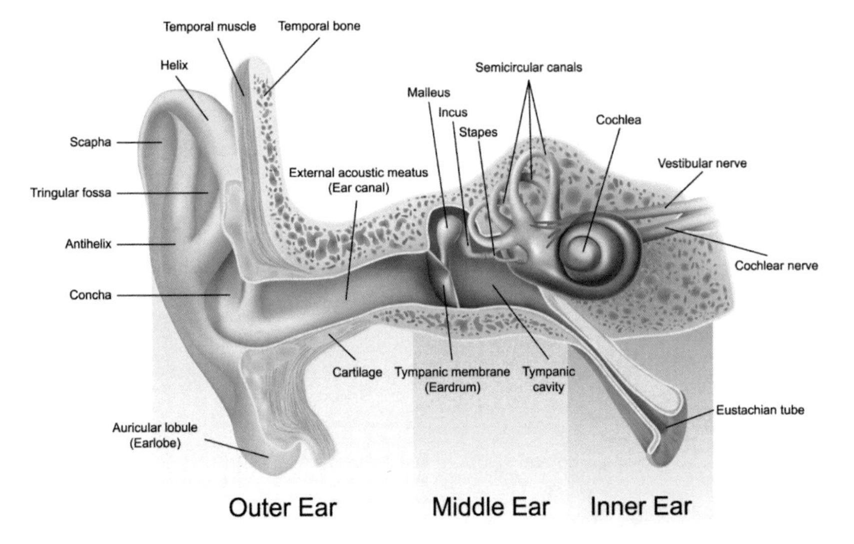

This diagram of the middle ear shows the tympanic membrane (eardrum) and the auditory (Eustachian) tube. The oval window, which is connected to the stapes, connects the middle ear to the inner ear. An imbalance of air pressure in the middle ear from external pressure changes on the tympanic membrane or severe pressure swings in the auditory tube may lead to rupture of either membrane, the tympanic membrane being significantly more vulnerable. *(Photo by Peter Junaidy/Adobe Stock)*

predispose an individual to middle-ear squeeze. Vertigo, hearing loss, and perforation of the tympanic membranes may result as well. Forceful attempts to equalize middle-ear barotraumas may create a pressure differential great enough to rupture the oval window that lies opposite from the tympanic membrane of the middle ear. The result is leakage of perilymph fluid, hearing loss, tinnitus (ringing in the ears), and vertigo.

The sinuses may also be involved in a process called "sinus squeeze." There are four pairs of sinuses that connect with the nose through tiny channels called ostia. In the setting of nasal congestion, the ostia may contract. As in middle-ear squeeze, the body becomes unable to equilibrate gas-filled spaces. Sudden pressure changes cause acute facial pain or even nosebleeds.

The least common of the "squeezes" is barodontalgia (tooth squeeze), pain or injury that affects teeth due to changes in pressure gradients during flying, diving, or hyperbaric oxygen therapy. Of 347 total cases of barotrauma reported in the 2008 edition of the *DAN Annual Diving Report*, only two cases of barodontalgia were listed, so the problem is uncommon in diving injuries. During the ascent from scuba diving, air can enter and become trapped behind an incomplete filling, loosening or even dislodging it due to a pressure difference when a gas-filled cavity doesn't communicate with the exterior and pressure cannot be equalized. The result is pain edema or vascular gas embolism.

These pressure changes may also cause fractures in teeth or dental restorations. Because tooth decay is found under fractured restorations, it is believed barodontalgia is caused by the movement of fluid from decayed dentine to the pulp of the tooth, the center of the tooth made up of living connective tissue and a subpopulation of dental pulp stem cells (DPSCs). Infection of this area (pulpitis) may become a serious problem.

The Fédération Dentaire Internationale (FDI) recommends an annual dental checkup for divers in addition to refraining from flying in unpressurized cabins or scuba diving within twenty-four hours of a dental treatment requiring anesthesia, as well as waiting seven days after an oral surgical procedure.

SELECTED BIBLIOGRAPHY

Pollock NW, Dunford RG, Denoble PJ, Dovenbarger JA, Caruso JL. Annual Diving Report, 2008 ed. Divers Alert Network: Durham, NC; 2008; 46.

Lynch JH, Bove AA. Diving medicine: A review of current evidence. J Am Board Fam Med. 2009 Jul-Aug; 22(4):399-407.

Stoetzer M, Kuelhorn C, Ruecker M, Ziebolz D, Gellrich NC, von See C. Pathophysiology of barodontalgia: a case report and review of the literature. Case Rep Dent. 2012:453415. Available at: http://dx.doi.org/10.1155/2012/453415

Acknowledgments

I am immensely grateful to Dr. Paul Phillips, Dr. Greg Adkisson, and Dr. Tom Neuman for their expertise and memories of David's case, which they shared with me. They are each remarkable individuals for the important roles they played in David's rescue and treatment as well as for their long and productive careers in medicine.

I was able to contact Dorothy Centa (later Dorothy Atkinson), who was also very helpful, being the only witness to several of the important events described in this book. I also met Dorothy's son, Shawn, who helped to clarify events that happened while David was unconscious in Mexico.

I thank Dr. James Holm, medical director of the Virginia Mason Center for Hyperbaric Medicine in Seattle, Washington, for his personal orientation to hyperbaric chambers and the physiology of hyperbaric medicine as well as his review of the technical sections contained in this work. Dr. Holm has been a scuba diver for more than 40 years and is a veteran of more than 4,000 dives. Scuba diving continues to be one of his passions.

Dr. Holm is also skilled in underwater photography, and one of his images graces the cover of this book.

Thanks to Lorraine Fico-White of Best Publishing, Leslie Hoy, and my wife, Pat, whose skill with punctuation, diction, and sentence clarity made this work better than I could have ever done myself.

About the Author

Monte Anderson was born in Kenosha, Wisconsin, and moved to Denver, Colorado, with his family when he was ten years old. After graduating from high school, he enlisted in the US Army, where he was first assigned to the infantry, then automotive repair, and finally, the medical corps.

After an unimpressive freshman year at the University of Colorado, where he spent most of his time on extracurricular activities like skiing, he entered the University of the Americas in Mexico City, where one of his professors sensed in him a talent for writing. Later, he worked for several years in pharmaceutical sales. From there, a desire to enter medical school was rekindled. During one of the most exciting times of his life, he completed required courses at the University of Nebraska at Omaha, then was admitted to the University of Nebraska Medical Center. He completed a medical residency at Creighton University and continued his studies with subspecialty training in gastroenterology and hepatology as an army officer at Fort Sam Houston in San Antonio, Texas. After his discharge from the military, most of his career was happily devoted to the Mayo Clinic in Arizona.

Feeling that true tales tend to be more compelling than fiction, he has always preferred reading nonfiction, especially since something is always learned in the process. *The Choice: A Story of Survival*, his first effort outside of scientific writing, is nonfiction.

Anderson lives in Prescott, Arizona, with his wife, Pat. They have three children, five grandchildren, and six great-grandchildren.